"Carroll is courageous in [sharing her] experiences. Her mandate is [to cut through fear generated] through media, disinformation and the truth embargo, all of which paints a very dark picture of our extraterrestrial ancestors, especially the Greys, or Zetas. Carroll describes a number of "intelligences" visiting us and shares her understanding of their nature and origins. Her book is a wonderful example of courage and honesty on her part, and it is a fascinating and insightful book that suggests that we face our fears, listen to our own resonance with such information, and explore with openness alternative perspectives as to who and what we are."
—Mary Rodwell, Principal of ACERN (Australian Close Encounter Resource Network) and co-Director of FREE (Foundation for Research into Extraterrestrial Encounters)

"Ms Carroll tackles many fascinating topics in the book. Our religious and spiritual beliefs are examined, the topic of individual and collective identity, our journey towards transcendence, and our very reason for being here on the planet. She is to be taken seriously. We dismiss her and her ideas at our peril. Judy Carroll's insights, her vulnerability, and her intellectual acuity are a force to be reckoned with. He or she who has an ear to hear, let them hear!"
—Rev. Michael J. S. Carter, author of *Alien Scriptures: Extraterrestrials In The Holy Bible.*

"I could not put it down. I read it all in one night, and the next morning I felt great! There is a Zeta energy in the book. They are subtle and loving, with a warm and fuzzy feeling. On a scale from cold 0 to hot-passion 10, they are a solid 5. Judy's book confirmed so many things that I had suspected. I very much enjoyed this book! It flows nicely, and the information is wonderful. It's nice to have someone else say that there is a lot of false 'spiritual' information out there. I have thought that for a long time."
—Jean Rockefeller, Healing Facilitator, Animal Communicator/Empath, Reiki Master, Shamanic Healer, Intuitive Liaison for Interspecies Communication.

"We commend this book to all who seek to know what is happening on our Earth and what the truths of the UFO matters are, and who seek to acquire an understanding of where the human race has been, where it is going and how we have been thoroughly misled upon the journey. It takes much courage to wake up and assess 3D life on Earth, as so much misinformation has been fed to us. And much of this has been based on fear. Judy Carroll has been fearless here, sharing both her journey and her learnings of life in other dimensions, and life with other life forms on other planets, principally the Zeta Greys, and also to some extent, the presences from Arcturus, Venus, the Sirians, Lyrans and Pleiadians. As Carroll tells us, the purpose of this book is to help us understand ourselves in terms of Oneness and Energy, thus enabling us to choose love over fear."
— Valerie Barrow, author of *Alcheringa*

About the Cover

Extraterrestrials and their craft have been depicted on rock walls and in paintings for as long as humans have learned to draw. These depictions come from all over the world so only a few are shown here.

Upper Left: Mother Mary with UFO in background. This is Medieval tapestry that hangs in Collegiale Notre Dame, Beaune, France.

Lower Right: Jesus's Baptism, which shows a clear flying disc in the picture, was painted in 1710 by Flemish artist Aert DeGelder. It is displayed at the Fitzwilliam Museum in Cambridge, England.

Middle: Person in a spacecraft is a portion from a 1350 painting The Crucifixion, which resides at the Visoki Decani Monastery in Kosovo, Yugoslavia, showing two spacecraft.

Upper Right: Native American Petroglyph at Sego Canyon, Utah dated to about 5500BC.

Lower Left: Aboriginal Petroglyph picture of ETs from St. Elizabeth Station, Australia

Extraterrestrial Presence on Earth

Lessons in History

Judy Carroll

Wild Flower Press
P.O. Box 1429
Columbus, NC 28722

©2018 by Judy Carroll

All rights reserved. No part of this book may be reproduced in any form or by any electronic or mechanical means including information and retrieval systems without prior permission from the publisher in writing.

Library of Congress Cataloging-in-Publication Data

Names: Carroll, Judy, 1952- author.
Title: Extraterrestrial presence on earth: lessons in history / Judy Carroll.
Description: 2 [edition]. | Columbus, NC: Granite Publishing, 2017. | Series: The Zeta series; 3 | Includes bibliographical references and index.
Identifiers: LCCN 2016059411 (print) | LCCN 2017008740 (e-book) | ISBN 9781893183650 (pbk.) | ISBN 9781893183667 ()
Subjects: LCSH: Grays (Extraterrestrial beings)
Classification: LCC BF2055.G73 C368 2017 (print) | LCC BF2055.G73 (e-book) |
DDC 001.942--dc23
LC record available at https://lccn.loc.gov/201605941

Manuscript Editors:
Pamela Meyer, Brian L. Crissey, Barbara White,
Manuscript & Cover Designer: Pamela Meyer

Printed in the USA.

Address all inquiries to:
Wild Flower Press, an imprint of Granite Publishing
P.O. Box 1429
Columbus, NC 28722
http://granite-planet.net

Foreword

Welcome, dear reader, to an intellectual treat and a spiritual test. Judy Carroll serves well, not only as scholar and teacher, but also as a spiritual guide. She provides her knowledge of Extraterrestrial (ET) civilizations in a scholarly manner, citing various sources and authors that confirm her personal experience. She not only informs but also teaches, comparing and evaluating various approaches to the understanding of ET civilizations and their impact on the history of Earth and Humanity.

Her service as spiritual guide adds to her contribution, as she not only provides information, but also questions and answers regarding the significance of the information. Also, she provides her readers with those procedures, which can assist them to evolve and to prepare for an age of enlightenment, when humanity moves from planetary persons to cosmic citizens.

She summarizes the writings of many authors and describes her connections with other souls — comrades in another lifetime who are here in this lifetime to continue the task of assisting Earth and humanity. Many "coincidences" or synchronicities are presented as signposts along her journey.

The author expands on the biblical accounts of the creation of Adam and Eve in the Garden of Eden. Various authors have written about channelled information and ancient scrolls regarding the creation of humanity. For example, Zecharia Sitchin has written several books about his translations of ancient scrolls. Krsanna Duran, author and publisher, has provided the second edition of her book, *Web of Life and Cosmos: Human and Bigfoot Star Ancestors*. Dr. Raymond Andrew Keller, PhD., Professor of History, West Virginia University, has provided a new book, *Venus Rising: A Concise History of the Second Planet*.

Foreword

These authors differ somewhat in regard to the specific ET civilizations who created modern humans on Earth, *e.g.,* Draconians, Pleiadians, Arcturians, Martians, Venusians, *et al.* However, there is general agreement that Atlantis and Lemuria were among the early civilizations, and there is general agreement that humans have a tradition of both patriarchal and non-patriarchal "gods."

Evidently, the ancient Egyptians had priests from both Atlantis and Lemuria. However, the male god, Jehovah, eventually was designated as the "only" god, and now, modern "Christians" are faced with a dilemma: is God male and judgmental? Or is God both feminine and masculine — and loving?

Dr. Keller, in *Venus Rising*, claims that Jesus (and Nikola Tesla) came from Venusian heritage, gentle and loving, not the Martian heritage of patriarchal fear and anger.

If that is true, then we can anticipate the possibility that true "Christians" can forgive our current planetary Controllers who are suppressing the knowledge of free energy *á la* Nikola Tesla, as well as suppressing the cancer-cure technology of Royal Rife. When free-energy devices are available, then we can envision a planet where every man, woman and child has access to clothing, food, shelter and education.

The author, as spiritual guide, shows again and again that the path of spiritual ascension is to minimize fear and to maximize love. Dr. David R. Hawkins, M.D., Ph.D., has shown in his several books, that the Map of Consciousness can be viewed as a chart for our various God views, as well as our various emotional reactions within the levels of consciousness.

Hannah Rosin, in *The End of Men* and *The Rise of Women*, has summarized various studies that show that women are emerging as leaders in academic, business, medical, military and political positions. (In the US, the only sections in which this is not yet true are in the U.S. Congress and among CEOs of major corporations.) If Rosin's conclusion is accurate, then soon the planetary leaders will be women, thus ending the patriarchy and the domination of men over women and children.

Other authors have focused on the studies of children: Crystal, Indigo, Rainbow and the Star Children. These writers claim that many souls now are coming into the planet — not only with high intelligence and high psychic abilities, but also with awareness of their roles in helping to heal Mother Earth and humanity and to effect a shift in consciousness.

Judy Carroll previews this shift from fear to love, and she previews the new players in the ongoing game of human evolution. Welcome, dear reader, to a new age of transformation. Enjoy the journey!

—R. Leo Sprinkle, Ph.D., Professor Emeritus,
Counseling Services, University of Wyoming

Publisher's Note

We live on a small, young, free-will planet in the backwaters of a small but ancient galaxy, which is itself but a grain of sand on a seemingly endless beach.

Our universe appears to be richly populated by highly developed, intelligent civilizations, so it should be no surprise to realize that Earth is not the only planet on which souls reincarnate. Souls also reincarnate on other planets, and sometimes souls who have lived on other planets inhabit human bodies on Earth. Judy Carroll is the current Earth-human name of a soul who, in her last incarnation, inhabited a small, gray-skinned, black-eyed Zeta soul container, or body, named El Or Kah. Her soul container was ruined when her scout spacecraft crashed in the 1940s, leading her and a number of other souls similarly bereft of physical form to incarnate in human form on this planet with us.

Judy's books, *The Zeta Message: Connecting All Beings in Oneness* and *Human by Day, Zeta by Night: A Dramatic Account of Greys Incarnating as Humans,* one non-fiction and the other a docu-drama based on a true story, form a prelude to this book.

This new book is the missing history of which we Earthlings intentionally have been deprived. Within that extraordinary history are the beautiful stories of how advanced civilizations have long been cultivating our physical and spiritual development, despite our ignorance and resistance. So, sit back and take it all in. Learn who you are and why you are here.

Brian L. Crissey, Ph.D.

Table of Contents

Foreword   v
Publisher's Note   ix

The Mission
1   Background on the Mission   1
2   New Paradigm Summary   6
3   Fellow Companions on the Mission   10

Beyond 3-D Reality
4   The Nature of God   16
5   Our Levels of Being   20
6   The Astral Plane   22
7   The Human Ladder and Evolution   25

Manipulation of Earth-Humans
8   Development of the Earth-Human Species   40
9   The Other Players   61
10   Repterran Controllers in Modern Society   73
11   Invasion of Free Will   95
12   Humanity's Call for Help   98

Development of Earth-Human Spirituality
13   Distorted Spiritual History on Earth   103
14   Ancient ET Evidence in Australia   107
15   The Three Levels of the Adam and Eve Parable   123
16   The Lord's Prayer from the Aramaic   126
17   The Misunderstood Concept of Original Sin   128

CONTENTS

Empower Yourself and Evolve
18   Science and Religion   136
19   Death and Dying   139
20   Reincarnation and Our Multidimensional Self   146
21   Being Free to Evolve   150
22   Getting Through the Fear Barrier   160

Our Future
23   Universal Spirituality   170
24   Interdimensional Emotions   176
25   Demonizing of the ET Visitors   181
26   Star Children   185
27   Understanding the Grey Guardians   191
28   Final Thoughts   207
     Glossary   215
     Resources for Further Research   223
     List of Figures   226
     Names and Works   227
     Index   232

The Mission

"If we're growing, we're always going to be out of our comfort zone."

— *John Maxwell*

"There will be help and support for those of you who work for higher consciousness. My people, the Star Elders, and interdimensional guides are willing to offer their protection. These include animal spirits and the souls of your ancestors who live in the Alter-Universe.
They also can hear your emotions and read your thoughts; they can join you and assist."
—*Kamooh, Sasquatch Elder*

1

Background on the Mission

My last life as the Zeta Reticulan, Lorkiah (pronounced El Or Kah) ended soon after the crash of one of our ships on Earth. I survived for some weeks before making a conscious transition from my physical body. I managed to retain full consciousness as I passed through the portal of "death" in order to be able to continue the mission of which I was, and still am, a part. After my transition and prior to reincarnating into the present life as Judy Carroll, further plans and decisions needed to be made in light of the deeper understanding that had been gained during my weeks of contact with Earth humans.

The reason for our coming here at that time was because Planet Earth has been firmly under the control of a rebel force over the span of many millennia. During this time several unsuccessful attempts at intervention had been made. However, matters really came to a head with the detonation of the bombs that ended WW II. Universal protocol generally prohibits off-planet intervention, unless a situation arises which threatens other planetary cultures and/or dimensions. This situation came into being with the development of nuclear weaponry by the military forces of Earth. It was time for serious and combined intervention, but reconnaissance and careful preparation was first required.

Our plan was to approach the authorities on Earth for discussion so that a warning could be issued on the dangers of such devices, but there were difficulties. Many Earth humans have been led to believe that they're the highest evolved life-form in the universe, and that no other human life on other planets exists. The only ones on Earth who acknowledged us were the tribal people from all over the planet. Stories of Creator Beings, Star Nations, Sky Fathers, *etc.* were still being passed down through the generations of Native Americans, Australian Aboriginals, African tribes, and many other indigenous cultures, so they were the ones we had always approached and did so again to make our presence known. Our ships were also being seen by others, so an "opening up" process was already in motion.

However, by meeting with and speaking to these tribal people we came to understand the depth of xenophobic fear that many Earth humans have of anyone who is "different." The Native Americans we visited told of the prejudice aimed at them by the non-indigenous society even before the world war. Now in the aftermath of such a major conflict we knew that it would be even worse. Those like me who survived crashes of our ships (caused by an interference to our control systems) got to experience this xenophobia first-hand.

There was no question that our Earth mission had to continue, so a between-life decision was made by all of us who had lost our physical lives in these crashes. We voted unanimously to reincarnate on Earth in Earth human form, coming "in the back door" so to speak, to try to make changes from the inside.

The problem we faced was that, because of the mission, we would need to retain our ET identity, but we also had to fit in fully as Earth humans. Some of us had never experienced an Earth life, and to be reborn as a native of the planet with our off-planet consciousness intact would make life extremely difficult. The solution was to retain our energy/soul connection with our previous culture—in my case the energy signature of Lorkiah's Zeta Grey consciousness—and to create an additional soul link with the Earth human Collective Consciousness. This would involve us incarnating as "blended" souls, resonating in energy synchronization with two rather than one planetary Collective Consciousness. Most souls select a single planetary culture to resonate with per life span. We were only allowed to do this doubling up because of the need to retain conscious awareness of our mission. We are not alone in doing this, however.

The other requirement was to allow the Earth human consciousness to "take the driver's seat" throughout our childhood. This would enable us to meld more easily into life down here, and to understand fully what it is to actually *be* an Earth human. Once this was accomplished, and when the time was right, we would then reawaken consciously to our other culture. We'd then be able to take up the mission from where we'd left off in the last life.

Energy can never be destroyed or created — it just changes. On the soul level of our being we are all eternal facets of Source (God) Consciousness. All souls are "created in the image of God," are therefore multidimensional and can operate at whatever level of awareness is required, including third-dimensional physicality. The 3-D body/vehicle, or "soul container" as the Greys call it, dies, but the real "us," the soul consciousness, is eternal. At the point of "death" we simply step out of the body. As a Zeta Grey I was able to do this in full consciousness. In this way I did not lose my full awareness as Lorkiah. In other words, I was able to retain full self-awareness. It was not an easy transition under the circumstances, but it was a very welcome one!

Throughout the whole of my present-life childhood as "Judy," an awareness of the unbroken flow of this other life remained, like a half-remembered dream on the very edge of my mind. Again and again I insisted to my bemused parents that - "I am *not* Judy! That is *not* my name!" As confusing as this was, valuable lessons were being learned, as my child mind wrestled with vague awareness of these strange "other beings" who visited me nightly, taking me out of my sleeping body to go on board the ship. The Earth human "Judy" part of me experienced deep fear of this strange contact, but at the same time, the "Lorkiah" part felt sad, lost and homesick. As hard as all of this was, it has enabled me to empathize fully with others going through the same process. I was also aware of strong guidance throughout my childhood and teenage years, with specialist training given to assist me to eventually take up the mission again.

By age 30 I was married and living out of town on an acreage property with my husband. On a sunny afternoon in March of 1983 a fully-conscious reunion with my Grey family took place. The story of this is told in full in my book, *The Zeta Message*. It involved a type of "walk-in" experience in which Lorkiah, the other "half" of my soul consciousness, was re-awakened to physicality in the Earth body to continue the mission, bringing with her a massive download of information and memories from the Grey Collective Consciousness. For this "walk-in" no near-death-experience was required, because the blended soul was already in place, so no soul energy needed to vacate the body to leave room for the other. It was all there, ready, willing and able. All that was needed was a deeper awakening process.

Detailed advice was given on a path of study I needed to undertake in order to carry out the required work. With his typical Grey sense of humor, my Elder/Teacher Maris, who was overseeing the process, remarked: "Okay Lorkiah, you've spent enough time hiding yourself away out here (on the country property). It's time for you to *get back to work!*"

And so, three years later, I found myself back in the city, following Maris's advice to learn meditation techniques in order to facilitate easier communication with "headquarters," and to learn Tai Chi and Reiki to help with the meditation and to keep the energy system in a state of balance and harmony. I was also told that eventually I'd be teaching these modalities to assist others to work consciously with energy.

Many synchronicities and amazing "coincidences" happened during the intervening three years to enable all of this training and preparation to take place. One of these was being part of the meditation group for eight years where I learned automatic writing. This was wonderful preparation for the books that the Greys asked me to write some years later to impart so much of their information. This was to counter the complaint made by many Earth humans that the Greys refuse to communicate. My role here is that of an ambassador between the two cultures. One of the Elders once commented that I'm having a "sabbatical" on Earth. Thus I became consciously aware of the double life I was leading, as an Earth human by day and a Zeta by night.

My first two books, *The Zeta Message* and *Human by Day, Zeta by Night,* were published in 2011. Just as I breathed a sigh of relief at having accomplished that part of the mission, in November of that same year I found myself in attendance at a meeting up on one of our ships. The purpose of this gathering was to discuss the progress of our intervention around the hijacking of Earth, which is the core of what our mission is about. You'll learn more about this hijacking shortly. Many other ET folks were there, and all of us were expected to participate and offer our ideas and services. Having only just gotten my two books out, I tried to hide my small, gray form behind a couple of tall Pleiadians. After all, I'd done my job! What more could I do? But hiding behind Pleiadians on the ship just doesn't work. "Hey, Lorkiah!" The telepathic summons engulfed me. "How about another book! People on Earth really do need to know how this hijacking problem began, who was involved, and even more importantly, the fact that it's continued to be a major block to the evolution of Earth plane humanity, and the very reason why we're carrying out this mission. Earth people also need to know what they need to do to free themselves. How about it? You can do it! The rest of the crew will be woken up to help out. That will be a very valuable contribution to our mission from all of you."

So, it took longer than expected, but here we are, here I am, and here it is – happy reading from all of us!

2

New Paradigm Summary

Human Development on Earth

- God is not a physical Being but rather a very high frequency Energy referred to by higher ET cultures as Source or Oneness
- The Earth-human race was created/developed by a number of off-planet groups known by several names – Assistant Creators, Elohim, Watchers and Guardians, who tap into God Source Energy in order to carry out the work of creation on many planets.
- On other planets throughout the universe, humans have been developed from many different animal forms, including mammals, insects, birds, reptiles, amphibians, *etc*. Many of these are more spiritually advanced than are the mammalian human species of Earth.
- Human form throughout the universe always follows the same basic pattern of walking upright on two legs, and having two arms, and a head, but there are many variations, depending upon which animal species a particular human planetary culture has evolved out of. Today's Earth humans were developed from a primate species, which is evident if you look at the skeletons of earlier humans.
- Human life on other planets also shares certain similarities with the animal form from which they've evolved – for example the Zetas' soul containers originated from insectoid stock, and they still retain the thin body and in some cases the chitinous exoskeleton and compound-eye structure of some insect species.
- Human evolution is not so much a physical process but rather a mind/spirit process of expanding and deepening conscious awareness. As humans, we create our own reality by the way we think, so learning to think in a positive rather than in a negative way is the key to ascension/evolution.

- There have been six genesis events on Earth - the first one resulted in the Fish/Mer-People, the second produced the Ant People, the third was the emergence of the Lizard people, the Fourth was the Bird People, the fifth was the Sasquatch People, and the sixth resulted in *Homo sapiens*, who evolved into modern-day humankind.
- Some members of this reptoid species that I call Repterrans or Controllers rebelled against the original assistant creator group and took control of Planet Earth. Wars were fought over this, but the Repterran Controllers and their Blond allies (Reptoid/Pleiadian hybrids) remain to this day, still in charge, having incarnated and reincarnated on Earth many times. They are also known as the Elites, Illuminati, or the Lower Lords.
- The Repterrans are determined to remain in control, keeping the younger human species (modern man) controlled and disempowered. It is for this reason they either officially deny or else demonize genuine ETs such as the Greys and others who are here to help humans evolve to higher levels of consciousness, which is the path of human evolution.
- No negative "aliens" are invading Earth. All negative interference is from right here, often disguised as "aliens," but really being carried out by the Repterrans and Blonds, who consider themselves to be the owners of Planet Earth. This group was originally from off-planet, is highly advanced technologically, has clandestine links with the military, church and other authorities, and operates both on the physical and astral planes of Earth. Great discernment is needed here.
- No negative beings from other Level-One planets can enter the energy field of Earth, because the energy fields of all Level-One planets stop such interference from happening. All negative interference is coming from either the physical or astral plane of this planet.
- Many "UFOs" that are seen, especially around underground military bases, belong to the Repterrans, and many negative "alien abductions," particularly those with black magic elements or a human military presence are Repterran-controlled rather than genuine ET encounters

New Paradigm Summary

- The Repterran Controllers are disempowering and interfering with peoples' energy bodies through many diverse channels such as putting out fear-orientated "channelled" information and predictions of gloom and doom, war-mongering, racism and religious intolerance. They have created religions based on fear and damnation, books, movies and video games that focus on violence, sex and fear-mongering, popular songs that promote feelings of aggression and anti-social behavior, materialism, consumerism, alcohol, tobacco and illegal drug industries, *etc.*
- The astral plane is confined to the Earth-human mind. It is formed out of the collective consciousness of humankind. It does not extend beyond this, and it is inhabited by many "thought forms," both positive and negative. It is manipulated by the Repterrans through their mind control of humanity. The "Akashic Records" of Earth are on the astral-plane level. Again, discernment is required when interpreting this information.
- Humanity is in the process of stepping up to a higher evolutionary level, but help is needed with this process, hence the presence of a large number of benevolent ET and interdimensional beings on Earth at this time. This assistance has been requested by many humans on deeper subconscious levels of their psyche. An automatic call went out after the first atomic bomb was detonated. This is why the ET presence has been stepped up greatly since World War II. They've actually been here for millennia, as evidenced by similar tribal lore from all corners of Earth that tells of visitations by Star Nations, Sky Fathers, Creator Beings, *etc.*, but visitation has now been increased because of the planetary shift.
- Fear is the limiting factor in human evolution, creating a "false ceiling" of perceived reality. If we wish to evolve as humans we must move past the fear barrier.
- Everything in the universe is energy and oneness. The only real separation is the rate of frequency.
- Many higher planetary and interdimensional helpers are here to assist Earth and human-kind's shift from Fourth- to Fifth-world consciousness. The expansion of human mind/awareness is the key to this evolutionary process.

- The Greys, or Grey Guardians have evolved from many different planetary cultures, are part of the Creator Race and specifically work in the field of energy transformation.
- A major challenge for ETs is trying to provide simple 3-D answers to questions that really demand multidimensional understanding.
- The Greys/Zetas think more in terms of Oneness, so are much more homogenized and amalgamated as a Group Soul Consciousness. Energy is always directed for the good of all, whereas Earth humans tend to be more segregated and divided.
- Greys/Zetas are capable of individual thoughts, personalities and ideas. They can choose to tune into the group consciousness or not.
- Some believe the Greys are an open book. We need courage on our side to open our eyes and minds and avoid the blinding influence of fear, and to accept a gift of love that they seem so willing to share.

3

Fellow Companions on the Mission

We all incarnate onto this planet (and others) for various reasons and into different bodies in different areas. Somehow we find our Soul Families and reconnect to do the work we need to do. The following women are some of mine who are part of the overall mission we chose in this life time. Through "coincidence" we have been drawn together to do this work. They all have contributed greatly to this book and my life.

MARCELLA BECKWITH
(SANNI CETO)

When I came across Sanni's first book, *Stranded on Earth: The Story of a Roswell Crash Survivor* some years ago, I immediately recognized her from the past Zeta life in which we knew each other. I contacted her and the recognition was mutual, with her referring to me by my correct Zeta name.

SANNI CETO

She retains a vast amount of conscious knowledge about her ET life, which she revealed in her second book, *Zeti Child: Lost Upon a One Star World*. Several researchers have questioned her in depth and were impressed by the answers she's provided. One of these researchers was Dr. R. Leo Sprinkle, Ph.D., for whom I've always had the greatest respect. He knows Sanni personally, and has often spoken with her at length over the years. In addition, as he wrote in her book, he has a great deal of respect for her.

Diagnosed with autism and recognized by the authorities down here as a Zeta hybrid, she was removed from her family and raised in foster care, which was not a pleasant experience for her, to say the least. Sanni's life has not been easy by any means, but nevertheless she has achieved much through her amazing artwork, her writing and her ongoing PR work for autism. She has also managed to retain her independence and dignity against massive odds. As she once said to me, "Earth is a school planet where valuable lessons can be learned. Prior to a soul incarnating on Earth, you choose the body you need for this incarnation as a way to teach yourself dignity and humility, and to learn what it is to be human. Autism has allowed me to connect consciously with all of nature — the trees, the rocks and the animals, all of which I love." Part of her mission here is to teach people about autism, which affects 1% of all people, and 1.5% in the US. Autism Spectrum Disorder (ASD) runs from low- to high-functioning ability. It's a hidden developmental disorder that affects every aspect of a person's life, causing the brain and nervous system to function differently. She helps people realize it is not a mental disorder but rather a neurological disorder.

JACQUELIN SMITH (ZAN TU KAI)

Jacquelin is a world-renowned animal and star being communicator, and a Certified Hypnosis Practitioner who assists experiencers with processing their encounters with star beings.

Jacquelin has authored two books, *Animal Communication: Our Sacred Connection,* and *Star Origins and Wisdom of Animals: Talks With Animal Souls.*

Her mission is to reveal the deeper aspects of animals to the humans who cannot hear them speak. Her animal communica-

JACQUELIN SMITH

tions fit perfectly with our memories of the work that we carried out as siblings in our past Zeta life together.

Jacquelin lives in the US and I live in Australia, and prior to our first Skype call, we'd never met (in this lifetime). We continue to find more and more in common.

HELENE KAYE (KA LI YAH)

HELENE KAY

I met Helene Kaye in 2000 via a mutual friend who knew of my ET connection. Helene's family had started having strange encounters in their home, much of it centered around their two children, Ben and Kira, aged 7 and 12 at the time who are now thriving adults. A full account is given of this contact in our non-fiction book *The Zeta Message*, co-authored by Helene.

As we got to know each better over time, we became aware of a very deep and ancient connection between us. I am not only grateful that she is here on Earth with me to carry out our mission together, but I also consider her to be my very dear friend and fellow star traveler, who was Ka Li Yah in our recent past life. Luckily, Helene is still around to navigate me through the difficulties of human life on Earth, and there is no way I'd be able to do this work without her. She is my dearest friend, guide, mentor and sister-in-spirit, with the most amazing capacity for love, compassion and humor. I might add that it's this last attribute that has gotten us into trouble, but also helped us through many trials and "sticky situations," both on Earth and elsewhere over the span of many lifetimes. Helene is a Reiki Master, a practitioner of Australian Bush Flower Essence therapy, and a qualified hypnotherapist trained in Dolores Cannon's QHHT technique, specializing in helping clients to process past-life regressions and ET contacts.

VALERIE BARROW (EGARINA)

VALERIE BARROW

I am very grateful for meeting Valerie Barrow (Egarina), but also for her insights into our past life together and confirmation of our mission here on Earth. She is also a fellow Australian author, who has provided major knowledge of Earth's history in her book, *Alcheringa: When the First Ancestors Were Created*, which tells the story of a group of these off-planet colonists whose star ship crashed on the east coast of Australia many millennia ago. This book, like mine, was also written from the extraterrestrial rather than the Earth-human perspective. In addition, it contains a huge amount of independent validation that has been provided by many people, and it presents the Genesis Story with respect and acknowledgement towards the Aboriginal people, whose ancient lore is based upon the off-planet Creator/Ancestor Beings who were responsible for the seeding of Earth and the development of the Earth-human species.

Valerie was led to write the book because she and her husband, in different soul containers, were among those colonists, and, in her present Earth life, she has been reassembling the group of reincarnated colonists who are once again on this planet to help the people in Earth-human form to evolve into a greater understanding.

Beyond 3-D Reality

"At the center of your being you have the answer;
you know who you are
and you know what you want."

— *Lao Tzu*

4

The Nature of God

"God" is not a physical being, but rather an omnipresent and omniscient Energy/Consciousness referred to by higher ET and interdimensional cultures as Source, Oneness or The Forever. Humans of Earth have created both God and the "angels" in their own image. God is not an old bearded man sitting up on a cloud doing the work of creation, but rather an Energy Source that just "is," who could be likened to an all-powerful sun at the very center of the universe whose rays penetrate and give life to the whole system. God does not "do" — God just "is" — and the Creator Beings, of which I'll also speak in connection with Earth's history, are able to consciously tap into this omnipresent and omniscient energy source in order to carry out the process of "doing," which is the creation, development and evolution of all life forms throughout this universe.

Many ET groups have visited Earth over millions of years in the role of Creator Beings, including Pleiadians, Arcturians, Sirians, Zeta Reticulans and Draconians, as well as many others. As channels for God/Source Energy, the Grey Guardian teachers with whom I've had a close association over many years have imparted information on God, energy, and deeper consciousness that I'd like to share. The three main things they've taught us are:

- The concept of Oneness — We are all One, Earth humans, Zetas, Pleiadians, Sirians, etc., and *all* life forms in the universe. We all share the same stream of consciousness, or universal life-force energy. There is no "them and us." In the greater reality we are all One.
- We are all immortal souls and an intrinsic part of God/Source/Oneness. We are *not* just a physical body of flesh and blood. The

physical body is just a vehicle — a container for the spirit/soul through which it can experience physicality for a lifetime.

- Acknowledging this understanding consciously enables us to tap into our *own* power within — our own God Center, which, in turn, enables us to break through the limiting barriers of fear to reach a state of pure, unconditional love.

How can we perceive things more deeply, seeing past the shallow outer layer of any given situation? The Elders tell us to always look for three or more reasons for anything. The ETs don't see reality in terms of physical matter. They see everything in terms of energy. Things are not always as they appear on the outside, but Earthlings perceive reality that way, limited by narrow 3-D awareness. The ETs are trying to help us move beyond this limitation, which can be very daunting for the human psyche.

> *God is not a physical being, but rather an omnipresent and omniscient Energy/Consciousness referred to by higher ET and interdimensional cultures as Source, Oneness or The Forever.*

Jacquelin Smith, star being communicator, is another "blended Zeta soul," and has been a contactee all of her life.

To give you a small example of their viewpoint, I have excerpted from her book *Star Origins* where she is interviewing a Creator Being who is an Overseer of Oceans and Water and gives one more perspective of the great reality in a way that we are not used to hearing.[1] Her questions are in italics.

Can you say more about assistant creators?

There are many assistant creators or co-creators who were part of Earth's creation. The Key Creator *is* the All, and created All. It created others as well as me to assist in creating other worlds and galaxies. The Key Creator's divinity and consciousness exist in everything, since we are all born from the Key Creator.

Does love enter into all of this?

---

1. pp. 105-6, *Star Origins and Wisdom of Animals*

The Nature of God

I hoped I had made myself clear, but let me try again. Yes, love! Earth was created so that a wide range of beings with various frequencies could come together to meet and bless Earth as well. The intention of creating all this was and is created from love and light.

Were there other beings who had other intentions?

"Yes! There were some who had other intentions. They were focused on power in regard to their creations. Some had more interest in controlling what was happening with their creations without thought to the beings they had created. Yet, there are those creators who serve the Key Creator and choose to step back and watch their creations learn, thrive and evolve. I serve the Light and oversee the waters in the best way that I can and will continue to do so."

What do you think or feel about how humans affect the waters?

"Humans have a choice whether or not to wake up and see that they are destroying this incredible creation and the beings who live in the waters. More humans are waking up to the fact that Earth's waters are living, breathing beings.

There are many beings in the oceans who have never been seen by humans. They swim and live in the deepest depths doing their work. Also, there are many beings in the oceans' other dimensions and galaxies who are doing a mighty work for Earth."

Why are they doing this work?

"Because they choose love and have chosen to be a part of this ongoing creation of Earth. They desire evolution for the All. It is that simple."

All life in the universe is "made in the image of God" in the sense that Source Energy permeates every level of creation from the largest galaxy down to the smallest part of every atom.

As Jesus once said: "My Father's house has many mansions!" (John 14:2).

Sasquatch Elder Kamooh, writes in *The Sasquatch Message to Humanity*, "We do acknowledge the existence and understand the presence of a Universal Creator Spirit, whom some of you may call

God, as the Unique Source and essence of all energy and consciousness in the Omniverse in all dimensional levels, that are ever diversifying into a greater variety of individual entities and forms of life, during the course of the eternally ongoing process of creation and evolution."

5

Our Levels of Being

Our soul essence consists of several interdimensional layers, all vibrating at different frequencies which combine together in harmony to resonate as our personal "energy signature." Our first 3 states of being are solid, liquid and gas, which make up the physical biological body. The fourth state is ether, and so, as multidimensional beings we have an etheric body, which vibrates at a higher and finer frequency than the solid, liquid and gaseous matter of the physical body. The etheric body is on the level of reality just beyond physical, and is the substance of our energy body/system in which the energy points known as "chakras" are located. This energy body permeates the physical body and also extends several inches outside of it. The etheric is the first, inner layer of our aura, which is made up of many layers of even higher states of being.

All Eastern medical and exercise traditions are based on maintaining balance and harmony in this underlying energy system, both as a preventative therapy — as all conditions of illness begin as an imbalance in this energy field — and also to address ongoing health problems by restoring balance to the energy system. This is how modalities such as Tai Chi, Qi Gong, Yoga, Reiki, Acupuncture, Shiatsu etc. work. Western medicine, which focuses more on treating the physical symptoms, is now beginning to acknowledge the benefits of these natural energy therapies, which deal with the *cause* rather than the *effect*. By treating only symptoms or the "effect," the cause is never touched and more havoc happens to the body. Ideally the two (physical and energetic) should work together in oneness rather than in competition with each other. Greys and other higher ETs combine both in their healing techniques, with great success.

Also like Tai Chi and Qi Gong, other martial arts such as Aikido, Kung Fu, Judo, Karate, *etc.*, make use of etheric energy for self-defense. The moves may appear to be only physical, but these systems are based on learning to strengthen and move etheric energy through the body using the mind, thus enhancing the practitioner's conscious awareness. Incredible feats of strength can be achieved, which are not done through physical strength but rather through the conscious cultivation of chi/ki/qi/etheric energy. The attack and defense moves are aimed at the opponent's chakras, and, when done in a certain way, can be lethal, or at least cause the opponent's energy field to be compromised, weakening them to a point where they fall to the ground.

You may also be surprised to learn that this technique is used by some Charismatic Christian "healers" who can, with the lightest touch or a wave of the hand through the recipient's energy field, cause them to drop spontaneously to the floor. This is not the "hand of God" or any sort of genuine healing, but rather a simple martial arts "energy trick."

6

The Astral Plane

The astral plane is a band of non-physical dimensional frequencies that was originally created as a "holding ground" for Earth-human consciousnesses while humans are out of physical form between lives. It doesn't exist for higher-dimensional beings, because once our conscious focus evolves to higher levels of the Human Ladder, we are able to consciously access these non-physical frequencies of energy, and our mind becomes much more fully aware rather than having parts hidden in sub-or super-consciousness.

The astral plane is composed entirely of the collective consciousness of Earth-plane humanity, along with myriad thought forms, both positive and negative, that have been created through all of the many belief systems of Earth. Because of the huge amount of disinformation that has been force-fed to Earth humans over the millennia and the huge amount of negativity that Earth humans have projected out, the astral plane is quite chaotic, with many conflicting thought forms. Hence the confusion on what is truth and what is fiction — what is real and what is not real. Like attracts like where energy is concerned, so the more fear that is built up in people's minds, the more negative, fear-oriented thought forms will be projected out and attracted back into their energy fields. In other words, the more fearful and/or superstitious you are, the more you will attract fear-oriented astral thought forms to you, which in essence you may be creating. Sometimes this can be mistaken for negative ET contact. Ultimately, continually thinking negative thoughts can be debilitating for people.

It is extremely important to understand that thinking positively has much deeper impact than just our earthly plane existence.

Astral Interference In ET Contact

Since the publication of my first two books, I have received communication from all over the world, from which it is very clear that there is a very high level of astral-plane interference on Earth. This interference is often mistaken for ET contact, because encounters with genuine ETs can potentially attract astral phenomena as well.

ET contact has the effect of removing certain negative blockages from the aura. However, a by-product is that a channel can be opened up. This is actually part and parcel of the spiritual awakening/evolutionary process, and could be looked upon as a test, or learning experience that sometimes occurs.

For instance, once a certain level of initiation has been attained, Eastern adepts and yogis must undergo a trial. The initiate is locked in a darkened room and comes under attack by lower astral entities, which must be overcome as part of the initiation process.

A person's own developing consciousness attracts these entities, because they sense the increased positive polarity being emitted. Energy is like water in that it seeks a balance. Therefore, a high level of positive energy draws a similar amount of negative energy. Negative lower-astral entities will try to feed off the higher positive energy because they sense the increased positive polarity being emitted, which is why poltergeist activity sometimes starts up around ET contactees. As with trainee yogis and adepts, this is a learning experience, or test. It often involves overcoming fear, but it can also be a test to see if the ego has been overcome.[1] Once you pass the test, such contact generally stops. This concept is also illustrated in Jesus's life, when he was "tempted by Satan" after he had been fasting in the desert.

---

1. A description of two such experiences that happened to Helene's daughter, Kira, are explained in *The Zeta Message*, on pages 106-109 and 150-154.

Over-inflated ego is in reality just another expression of fear. If an Earth human operates through inflated ego, then that is the sort of astral thought form that will be attracted to them. This in turn feeds their ego further. This is the reason why psychics and channelers need to be extremely careful of their motives. If there is any fear, ego, self-righteousness, greed or manipulative tendencies present, then that is the sort of "guide" they will draw to themselves.

Higher (beyond astral plane) beings have no ego, and are not caught up in titles, labels and the like. Generally they do not label themselves as being a "master," "high priest," or "angel." Their messages are always positive, never fear-oriented or controlling. The only emotion behind such messages is unconditional love, nothing less.

To summarize, the astral plane is the plane of dreams, illusion and emotion, and can potentially cause major blocks to evolution on the soul level. Astral-plane entities and thought forms can be drawn into three-dimensional reality by human minds caught up in artificially induced fear, egotism and any other vices, because the astral plane is connected to and part of the collective consciousness of Earth-plane humanity.

Moving beyond fear and ego by recognizing and learning to tap into your inner Soul Essence and/or your God Spark is the key to freeing yourself from the illusions of the astral. Part of this too is being able to see beyond your "labels" of nationality, religion, planetary culture, *etc.* In reality these are nothing more than roles we play as part of our physical existence, just as an actor takes on a role in a drama production. Our true Self is beyond such limits.

7

The Human Ladder and Evolution

In 2001, Helene Kaye (co-author of *The Zeta Message*), her daughter, Kira, and myself were given a very important teaching and I feel it is essential to share this with you before we begin since it simply explains how evolution effects *all* beings in the Universe.

A tall Grey Elder Teacher whom we know as Oris explained that the Human Ladder is a path of evolution followed by all members of the human kingdom. This expands through ten levels of conscious awareness from "animal human" to "cosmic human." This evolutionary journey actually begins even further back, at the mineral-kingdom level, proceeds through the vegetable and animal kingdoms, and on to humans and beyond.

To understand the Human Ladder, you must understand that it is *not* a physical hierarchy. It involves *only* energy vibrating at ascending frequencies. The separation is purely energetic and it has nothing to do with "good" or "bad." It is simply different frequency bands, most of which are beyond the third dimension, because they are multidimensional. These frequencies also represent the evolutionary journey of each individual soul as well as planetary cultures. Because we are multidimensional beings, every one of us has a part of our being manifesting on each of the ten levels of the Human Ladder. Our soul *actually spans* the entire Ladder.

Earth-human conscious-mind focus is on the very first rung of the Human Ladder. In the physical universe there are about 1,000 other human species focused on this level, all operating at varying degrees of free will and on more or less the same level of conscious awareness. Regarding free will, there is actually not as much of it on Earth as people like to think. Most Earth humans have been very effectively and subtly programmed and brainwashed by the

The Human Ladder and Evolution

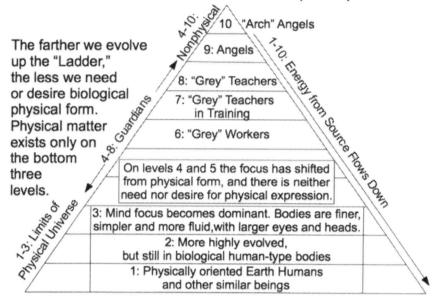

ILLUSTRATION AND EXPLANATION OF THE HUMAN LADDER

Controllers (who you'll meet shortly) through various religions and other belief systems. Because level-one humans are not in full control of their free will, most are confined to their own home worlds.

The frequency rate of the planetary energy forms a protective and confining barrier. More evolved (higher-vibrational) beings can enter a planetary zone of lower frequency, but less evolved (lower vibrational) beings cannot enter a planetary zone of higher frequency.

By the time human minds awaken to Level Two of the Human Ladder, their awareness has expanded enough for them to know consciously that "they" are not just a physical body but rather a spiritual entity. This soul essence has freely chosen to inhabit a physical form in order to undergo certain learning experiences. Once this concept is clearly understood and the ability to tap into a more expanded level of conscious awareness has been mastered, the entity is then able to freely leave the physical body at will. In reality, everyone leaves their body during the sleep state, experiencing it as dreaming. However, once a human can access Level-two consciousness, this body-leaving can be done easily, at will, and in full conscious awareness.

Many planetary cultures and highly developed Earth humans have reached this point. These are more "evolved" humans who have moved upwards from Level-One planets like Earth — planets on which they've had all their "rough edges" rounded off, so to speak. They've learned the value and importance of keeping their energies well balanced and harmonized. Because they can access more of their potential conscious awareness and have more DNA activated, they are geniuses compared to average Earth humans and are able to easily tap into fourth-and fifth-dimensional reality.

These beings— advanced mystics, shamans, *etc.* — are still centered in their biological bodies, but their deeper conscious awareness gives them greater control over their being's physical aspect. Hints of this can be seen in some of the Eastern adepts, who have mastered the ability to control body temperature, blood pressure, *etc.* In Level-two reality, illness and physical degeneration are less of an issue, hence their much longer life spans. Level-two civilizations have also moved beyond the Earth-human state of disharmony and fear and are much more peaceful. Because they are more attuned to Oneness, they lack any warlike tendencies. They have a much deeper respect for others and the environment, and understand how to access clean, natural energy. Consequently their technology is more efficient and far less destructive than here on Earth. Everyone's needs are met without poverty, hunger or misery. Life may not be perfect, but it's certainly an improvement from life at Level One.

When Level Three of the Human Ladder is attained, with 30% conscious awareness and active DNA, the need and desire for physicality begins to drop away. One's energy is focused much more on mind and spirit. At Level Three, the concept that we are far more than just a physical being is clearly understood. Level-three humans know that the physical body is simply a temporary "container," or "vehicle" to be used only when necessary and easily stepped out of to enable the pure soul essence to manifest free of the grossness of physical form. Such liberated soul essences can appear as orbs, or light bodies.

VISITING SOUL-ESSENCE ORBS IN JUDY'S BACK YARD

It's difficult to label any particular planetary group as being from a certain level, because many star systems have a number of inhabited planets of which some are focused on one level and others focused on another. For example, in the Zeta Reticulan system, some operate at Level-Two consciousness, while others are at Levels Three, Four or even higher. And remember, at some capacity we are at all levels.

Level Four and upwards on the Ladder is not physical as we know physicality on Earth. This is the first step into what could be called the Guardian realm. From this point on, the soul essence is able to consciously access higher and higher frequencies of reality, and eventually meld back fully into the Oneness of Source. As the entity's consciousness evolves upwards through these levels, it is moving from human to "angel" — from Earth human to cosmic human. What Earth humans perceive as "angels" are simply very highly evolved soul essences operating right up at the top of the Human Ladder. They have moved way beyond the need, or even desire, for physicality, and are more than 90% consciously aware. They are almost or completely pure mind/spirit. However, they can take on a semblance of physical form if they wish to make contact with beings on the lower frequency rungs of the Ladder. In reality, only a small part of the universe vibrates at gross matter levels like the Earth — most of the universe is beyond physicality.

These higher levels of the Human Ladder are the basis of various Earth-plane belief systems that speak of spiritual or angelic hierarchies that exist in the universe. These "hierarchies" are based on ascending and expanding levels of conscious awareness

THE CROP CIRCLE GIFT

It was obvious to us that the Human Ladder is very complex, and I really wanted a visual, less hierarchical, illustration to be included in *The Zeta Message*, which I was writing at the time. Both Kira and I tried again and again to draw it, as a series of spiraling "arms" radiating out from a central point and somehow coming back in to join up at the "Source" point in the middle. There were other smaller circles intermingled between and sort of linking it all together, but the complexity beat us. In the end, in sheer desperation and half-jokingly, we put out a request to our Grey family "Upstairs" to please provide us with a crop circle to be used as an illustration in our book to enable us to get this highly complex but extremely important teaching across to others.

A couple of weeks later, in mid-August, a report came on after the TV News broadcast, of the latest crop circle that had appeared overnight on Milk Hill field in Wiltshire, England. To be in the news, it had to be special and it was — huge— about 1000 feet across with 409 perfect circles. To make it more difficult to explain away as a prank, it had rained all night, but there were no footprints that morning.

The Human Ladder and Evolution

I was thrilled beyond words, because I'd visited this area several years earlier. As soon as it came up on the screen, I knew that it was our Human Ladder, being very close in shape to the illustrations we'd tried to do of it ourselves. I hurriedly loaded a video so we could have it on record to study further, and a few days later Oris confirmed that this crop circle had indeed been provided as an illustration of the Human Ladder.

MILK HILL FORMATION 8.13.2001

The Human Ladder has ten levels, but the 2001 Milk Hill crop circle only has six spiraling "arms" that represent the "galactic levels." Ever since 2001 I've been trying to figure this out, but it's only now, as I've been writing this book, that the answer has been given, and of course, as with all of this deeper knowledge, it's been "staring me in the face" all along.

I did ask Maris about this at the time, and he gave the rather lame excuse that: "Yes, as you found out for yourselves, it is very complex and hard to illustrate, even for us! We could only manage to show six arms." Knowing the Greys' amazing illustrative and energy abilities, this answer didn't really satisfy me, but often there is deeper information involved that we may not be quite ready to access at the time of asking.

Sure enough, it turned out to be the case. Some contactees get upset about not getting direct answers, and they accuse the Greys of being dishonest, evasive and non-communicative, but as our Grey Teachers have told us on more than one occasion: "You have all your answers within — you just need to dig deeper!" They refuse to "spoon-feed" us, knowing very well that finding our own answers and learning to trust our inner voice is an important part of our spiritual maturing and Self-Realization process.

I pursued the matter with my Teacher, Maris, and managed to grasp the fact that what Oris had referred to as "galactic levels" were not physical places, as such, but rather ten ascending and expanding levels of conscious awareness — in other words, a path or cycle of evolution, from animal human to cosmic human.

At the time they also spoke to us of "living ten lives on each level," which did not make sense to me, as we live many lives on each level. I now realize that, even within the information given to us on the Human Ladder, three layers of meaning can be found. At first I was only able to understand the first two — that it's a path of evolution on both the physical and the mind level of our being. However, it encompasses not only body and mind but also Soul/Oneness as well, so it also teaches us that we are experiencing lives on all ten levels of the Human Ladder simultaneously. The third, deeper layer of understanding that Oris was trying to get across to us is that we live a life on *all ten levels, all at the same time.*

The point to remember here is that our soul operates in Oneness, which is the omnipresent, omniscient God aspect of our being. In other words, all the knowledge, wisdom and understanding that we require is within us. God is not a separate Being sitting up somewhere on a cloud in "heaven." God is within each and every one of us — the still, small voice within.

The Human Ladder and Evolution

MILK HILL FORMATION CREATED IN THE ABORIGINAL STYLE

The Human Ladder does have ten levels, but it's only those beings in residence on the first six levels that are able to fully express physicality. The first three levels are completely physical. But you'll recall that even those on levels 4, 5 and 6 who have evolved beyond the need to experience lives on a physical planet can, by lowering their vibrational frequency rate, experience a physical incarnation on a planet such as Earth, as a number of star people are doing at this time. However, once the soul consciousness has opened up to full awareness of levels seven through ten, it is vibrating at too high a frequency to be able to experience a whole physical lifetime on a lower energy planet. Sure, there are aspects of the soul on all ten levels, but the Level-One aspect cannot access the full conscious awareness of its higher aspects while in physical form, and the higher aspect cannot experience complete physicality on the lower levels.

To illustrate this I'll use myself as an example. At those times when I link in to my Grey Teacher, Maris, who is on a much higher level of the Human Ladder, I'm aware of a massive amount of knowledge and information potentially available, but my limited Earth-human physically attached mind simply cannot process it. Even El Or Kah, my Zeta aspect, can't fully access the vastly expanded consciousness of a Level-Eight Elder. I know that Maris is an aspect of me that exists at a higher frequency, and that he can express part of his consciousness through me, but he could not exist down here in his own right for the full span of a lifetime, just as I cannot express his full conscious awareness through my physical

Earth-human or lower level Zeta brain. He can make a short appearance down here, but cannot remain for any length of time.

So, this being the case, the concept is very cleverly expressed in the Milk Hill Field illustration of the Human Ladder, with the first three physical levels and the next three potentially physical levels being represented in third dimensional reality through the six spiraling arms. However, the other four higher-dimensional levels are there as well, right alongside the physical ones, but vibrating at a higher frequency that cannot be perceived by our limited third dimensional human awareness. As in the case of ET script, the extra four "arms" are superimposed over the six visible three-dimensional ones. ET script is different from Earth-human script in that it spans multiple dimensions, and therefore must be read with deeper sensing than just physical sight. In most cases, crop circles are multidimensional but we see only the flat 3-D perspective.

A physical hint of these deeper layers is given in the small circles that can be seen alongside the bigger spirals. There are a total of 409 circles, which numerologically add up to 13, symbolically representing Transformation, which is precisely what the Human Ladder is all about, that is, transformation from animal human to cosmic human. The 13 in turn can be converted to a 4, the "built-to-last" number of the "worker bees" of the Universe.

Healing Aspects of the Glyph

How this glyph works as a healing mandala is through the fact that it illustrates the path of human evolution back to Oneness/Source and is a blueprint for human self-realization. It's the energy, or resonance of Oneness intrinsically contained within the circle which impacts upon the energy systems of humanity and Planet Earth, to assist with the realignment/ascension process.

Maris has asked me to include a statement here on behalf of the assistant creators in reference to the crop circles — "The crop symbols are our freely bestowed gift to humanity and your Mother Earth. They are what are known as True or Universal Symbols, and they invoke and evoke a high-frequency energy that links directly into your consciousness to assist in the ascension of both planet and people. Each glyph has a unique meaning for each individual because of the fact that they are Universal Symbols and therefore carry the omnipresent and omniscient Energy of Oneness.

"This energy is freely available to anyone who, through their own free will, chooses to tune into it. All you need do is select the pattern/symbol that resonates most deeply in your heart, and take a little time each day to meditate upon it, link with it, and clearly state your need and/or intention. If this is positive and in line with your soul destiny, it will come about!"[1]

Evolving Through the Ladder

As a human evolves through the process of reincarnation, which is an intrinsic part of evolution, their conscious awareness deepens, expands and strengthens. On Level One, a human can access approximately 10% of their potential *conscious* awareness. This refers not to physical brain but rather to mind/consciousness. Sub-and super-consciousness are not usually accessible to the average Earth human. In conjunction, around the same percentage of their DNA is active. The remaining parts of the DNA are what Earth scientists refer to as "junk DNA," which is far from the truth.

Quantum physics is now slowly changing scientific minds on this, particularly regarding the ability of human minds to create the reality in which we live.[2]

As Leo Sprinkle stated, "For thousands of years, mystics and teachers have been saying that this physical world is an illusion. For example, Jesus was quoted as saying, 'The Kingdom of Heaven is within.' His comments have been used to emphasize what physicists are (now) saying: Be in the world, but not of the world; behave like a human, in a world of perceived reality, but know that you are a soul in a universe of consciousness.

---

1. See Not Made By Hands in the "For More Research" in the back of the book.
2. See Dr. R. Leo Sprinkle's presentation "ET Experiencers: From Planetary Persons to Cosmic Citizens?" or see *Spontaneous Evolution* by Dr. Bruce Lipton.

"In summary, New Science is telling us that the Old Science model is not correct; perceived reality is not true. We are not observers; we are participants. Our observations, our perceptions, our intentions, exert a slight but measurable influence on the world within us, as well as the world around us."[1]

The higher the percentage of our "junk" DNA we awaken through further evolution, the more we will be able to influence the world in full consciousness. Remember that Level One of the Human Ladder is the most polarized in duality. As human-mind focus evolves up to Level Two of the ten levels, not only do people begin tapping into 20% of their potential consciousness, but also more of their DNA, which is what is happening on Earth right now with the "Star Children" since they carry a higher percentage of ET genetics. This increasing mind focus is also what the planetary shift is all about. The planet herself is stepping up to a higher-vibrational frequency, from the Fourth World to the Fifth World, as has been predicted by many tribal peoples the world over.

Fourth-world frequency operates in three-dimensional consciousness with subconscious access to the fourth dimension, wherein lies our personal energy body/chakra system. Fifth-World frequency will enable Earth-humans to operate consciously in fourth dimensional reality, with subconscious access to the fifth dimension, which will give humans much more conscious control over their health and physical well-being, and clearer access to higher astral-plane frequencies, including memories of past lives and spiritual helpers. And remember - parts of us are already at these higher levels, we just have to tune into those aspects of ourselves.

Dr. R. Leo Sprinkle continues with this theme on p. 17: "Gregg Braden, in... *Fractal Time*, offers a time code, which can be used to assess three levels of events: human, Earth, and cosmic levels. The end of a 5,125-year cycle, and the end of a 25,625-year cycle, provides humanity with the opportunity to 'choose' the next stage of our evolutionary journey. Perhaps, as suggested by Dr. Eric Pearl, we are

---

1. *ET Experiencers*, Sprinkle, p. 12

learning how to reconnect with our true selves; we are engaged in 'evolutionary healing.'"

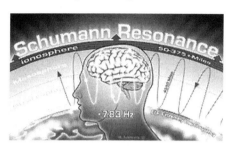

ILLUSTRATION OF SCHUMANN RESONANCE

The Schumann Resonance,[1] Earth's base-energy frequency, can be thought of as the planet's heartbeat and could be revealing Earth's evolution as well. For decades it remained constant at 7.8 cycles per second (Hz). It was so constant that global military communications were developed at this frequency. Recent reports claim that it has risen to 11 Hz and that the change has occurred very rapidly over about 20 years.

I am not alone in thinking this upward shift may be part of a planetary shift. Through his work, Gregg Braden shares scientific proof that the Earth is passing through the Photon Belt (the reality behind the misinterpreted Planet X, or Nibiru, fear-oriented myth) and the Earth's slowing rotation. According to Braden, when the Earth's resonant frequency reaches 13 Hz, we will be at a zero-point magnetic field, which, in two or three days, will produce a reversal in the Earth's magnetic fields.

The Schumann Resonance is intrinsically linked to the Earth-human energy system, since we cannot be separated from Earth. When astronauts are separated, NASA has had to install equipment on board their off-planet space craft to maintain this atmospheric resonance for the health of the crew. Bearing this in mind, any change in the vibrational frequency of Earth is going to have an impact on humanity.

---

1. bit.ly/1YJQmQs

SUMMARY

Evolution up the Human Ladder is really an "awakening of self" process. It involves awakening yourself to the point where you are able to consciously acknowledge all of these higher levels of awareness within yourself — in other words, self-realization.

Bearing this in mind, genuine contact with ETs also involves meeting up with various parts of yourself, hence the incredibly deep sense of recognition, soul connectedness and Oneness that is sometimes experienced in ET encounters. At the same time, there are also parts of self that need to be healed of deeply buried fears, so contact in these cases can be extremely traumatic.

Spiritual evolution up the Human Ladder entails exposing our fear/ego, which can be buried and hidden away on very deep levels of the psyche, built up as "excess baggage" over the span of many lives. This limiting factor must be cleansed and "detoxed" in order to enable spiritual growth to take place. This is where the so-called "Greys" may be required to lend their assistance, as "triers" and "testers" of souls.

The Greys' eyes have the effect of mirroring/reflecting back fears (often from past lives) that are buried in the sub-conscious recesses of the human mind. If a person wants to evolve spiritually, such fears need to be brought to the surface to be healed and cleared. If this is not done, the evolutionary path becomes blocked with the lower-frequency energy generated by the fears. Many people cannot do this on their own, so they make a pre-birth agreement with the Greys for assistance. Some people experience discomfort with the eyes of the Greys because of this fear factor, especially in the beginning of contact. At some point the eyes awaken a deep soul connection of unconditional love.

Because of the limited nature of Earth human conscious awareness, a lot of this clearing process must take place on the astral plane. This is the highest-vibrational frequency band that many human minds are able to access. The astral plane itself acts like a barrier and a mirror, creating and reflecting delusions of grandeur or fear, depending upon the soul "history" of the person.

In the same way, coming down from higher frequencies, the ETs and interdimensionals have trouble accessing the lower frequencies of the physical Earth plane. Therefore, the astral plane is used as a common meeting ground for these helpers to connect with the human who has chosen to evolve. Even though the contact is positive, the contactee's own fear or ego issues can cause problems. They perceive the experience as being frightening or negative, when in reality it is actually healing.

In the greater reality all of us are multidimensional. Every one of us has an aspect of our being on every level of the Human Ladder. Part of our awakening/self-realization process as humans is to gradually expand our consciousness to the point where we can consciously acknowledge all these different facets of our being as part of Oneness. In this way, the Human Ladder is an illustration of our Soul/Higher Self. In many instances, our spirit guides and ET visitors are actually *facets of our own Higher Self* existing at these higher levels.

Manipulation of Earth Humans

*"Some seekers will do anything for self-realization,
except work for it."*

— Sri Chinmoy

8

Development of the Earth-Human Species

Sasquatch Elder Kamooh: "As Mother Earth was being born, Her Soul was carrying within Her womb the souls of Her children. She was born from the Father Sun, who acted as a star gate, providing passage for substance and souls. This is how most planets are created, to allow sentient life to incarnate and consciousness to evolve. From their original ethereal entity and energetic reservoir of plasma, their soul and bodies manifest."

A number of different planetary and interdimensional groups took part in the creation and development of Earth as an inhabited planet. This process has been a combination of natural and assisted evolution, and these various assistant-creator groups have come to be known on Earth by a number of different names such as Guardians, Watchers and Elohim. The word "Elohim" translates as the plural of God, that is, gods. These more highly evolved beings certainly are agents of creation in that they tap into creative God/Source Energy in order to carry out their work, which involves the creation, development and evolution of life forms on the lower levels of the Human Ladder. The Elohim are basically cosmic scientists and explorers, but they are not "Key" Creator, which is God/Source.

These Elohim began the work of creation on Earth billions of years ago. They assisted in the terraforming of the planet, then initially sowed the seeds of life here. They came and went, sometimes returning after short periods of time and sometimes not for many millennia. Earth is a very out-of-the-way planet on the edge of the great Milky Way galaxy, and the Biblical story of Earth being created "in six days" is allegorical, with three levels of meaning. It refers to six stages in the terraforming of the planet Herself, six stages of the emergence of life forms originally from the ocean, and also to the six human genesis events that have taken place. This process has spanned billions of years, with mammalian Earth humans being the end result.

HUMAN LIFE ON EARTH

The history of Earth from its very beginnings is told in detail by Sasquatch Elder Kamooh. However, this book deals more with the Repterran Controllers of this planet and their ongoing relationship with modern-day humans. Between each of the six genesis events on Earth there have been major cataclysms, causing most of the populations involved to move on, either into other dimensions or to other planets. The first genesis, which resulted in the development of the Fish People, occurred eons ago when the Earth was entirely covered by oceans. These beings are recalled in myths and legends as the Mer-people, who still make occasional appearances in 3-D "reality."

The second genesis, as dry land emerged from the oceans, created the Ant people, who are also referred to in the Indian scripture the *Mahabharata*. They are ancestors of the Greys. I'll talk more about them under that heading later in this book.

The third genesis, which is where my book begins, was the emergence of the Lizard people, on whom I'll expand shortly. The scales of some Lizard groups evolved into feathers, resulting in the fourth genesis that produced the Bird people, some of whom have been mistaken for "angels." A rebel group of Bird people mixed their genetics with the reptoids.

Sasquatch Elder Kamooh: "The Bird People's kings who seceded from the Star Council to join with the Archon lower lords are known as the Anunnaki, Sky People fallen, or Rakshasa. From Watchers or Hakamin, they became Fallen Ones or Nephilim. They were exiled from Earth by the Star Council and went to live on their artificial planet Nibiru, abode of the gods."

The fifth genesis, and first mammalian human species, was the Sasquatch, who were removed from the Outer Earth and taken to another dimension in the Inner Earth by the Star Elders, to keep them safe from interference by the Repterrans. Unlike the sixth genesis *Homo sapiens* who have been blocked, the Sasquatch have been able to evolve naturally, and have developed a much deeper and broader level of conscious awareness that enables them to move freely and consciously between the dimensions and to work alongside the Star people. They can appear on the surface of the planet, but being highly telepathic and multidimensional, they can make a quick escape through the dimensions when they need to.

The development of all these various humanoids has involved the introduction of off-planet DNA. Human life forms do not evolve naturally, but must be bio-engineered.

Sasquatch Elder Kamooh: "The Star Elders in the (Cosmic) Council are the ones who have kept collaborating into inter-species, peaceful and spiritual relations, to maintain Cosmic Order. Yet, some were disqualified and expelled from it.

"Among them are the ones we refer to as the lower lords. They were Star Elders who had reached immortality but decided to regress into egoistic consciousness to impose control and possess power. They are also known as Elioud, the cast-out gods, fallen angels, demon lords, devils and dark masters. Their high level of scientific and spiritual understanding provide them with immense psychic powers.

"Over the aeons, they gathered one third of all worlds under their control, including this planet Earth. The ones who came onto Earth in the early phases of Her development were from a rebellious group of over 200 planets that had been developed and colonized by reptoid species of Draconians.

"They are called Draconians, not only because of their sizes and appearances, but also for their origins based in a star system you know as Draco. Since they colonized and controlled many planets in several star systems, the species and powers under their control cannot all be called Draconians, hence the term *lower lords* that we use. They evolved from reptoid genetics and kept using them to devise many hybrids.

"When the reptoid lower lords, under the control of the Draconians, engineered hybrids on Earth from reptile genetics, they devised the class of hybrids known as reptilians (whom I call Repterrans), to bypass and control the Lizard People, also reptilian hybrids who were created by the Star Elders in a higher Soul evolution.

"At this point you will realize that the story of every species or hybrid is far more complex than it seems. Therefore, we can't generalize or judge any of them with our limited understanding of their vast reality. We can only try to understand their role and purpose in our collective spiritual evolution and healing. Every soul and entity is always offered the possibility to change their way, heal and evolve, *as long as their consciousness doesn't regress into eternal lethargy, which is spiritual death.*" (My italics — Judy).

The bio-engineering of the Earth-Human (Repterran) species, the De-raks, the Te-raks and the Puy-yats, began during the time of the dinosaurs. The biological container to be developed had to be suitable for the prevailing planetary conditions. Genetic material was taken from the more highly evolved stock on Earth at that time, which was a bipedal reptilian. The Draconians carrying out this work were reptoid humans who were highly advanced in the field of genetic engineering. Modern humans still retain a reptilian part in their brains that is inherited from these first reptilian/human biological forms.

During earlier times on Earth, going back *billions* of years, all of the off-planet people carrying out genetic work on Earth were able to *consciously* step out of their biological bodies. However, because of their controlling and warlike natures, competition arose between the Draconians and their hybridized offspring over the "ownership" of the planet. The Te-raks developed a brain parasite that had the effect of interfering with DNA, which in turn caused the vibrational frequency of those affected to drop down the scale, which, in turn, caused the already fallen races to become even more caught up in the pleasures and temptations of biological existence. Thus their focus became more and more enmeshed in physicality and the need for dominion over others.

Development of the Earth-Human Species

As the Age of Dinosaurs passed, work had already begun on the development of a *mammalian* human species native to the planet and more suited to the changing climatic conditions. The base stock used was a type of bipedal primate. The main reason for the development of this new human species was to create a part-human slave race to carry out hard labor for the Draconians and Repterrans. These large and physically strong "mixture" creatures were only allowed to evolve to a point where they could understand and follow orders obediently.

They were blocked from further evolution by means of the DNA-damaging parasite so that they could be exploited to the fullest. This program was based in what is now the Mediterranean regions, the Middle East and Africa. Hence the discovery of "Lucy," believed by some archaeologists to be the original "Eve." A number of different genetic lines were experimented with in various parts of the world, resulting in ancient human primate-type remains being discovered in many countries. It was one of these ancient human primate groups who, with later off-planet (Pleiadian and other) DNA added, was taken to a safer dimension and evolved into the Sasquatch people. The Sasquatch people were created 50 *million years* before *Homo sapiens*.

By this time Planet Earth had become a violent and aggressive environment, completely under the control of the various reptoid humans, all contending among themselves for dominion. Earth was avoided by other off-planet people, all of whom had been driven away by the aggression and lower-vibrational frequency of the fallen species.

Pleiadian Rescue In Grey Bodies

This situation continued for millennia, but finally it became so bad that the off-planet assistant creators decided they'd have to intervene, as Cosmic Law was being blatantly disregarded. The story of this intervention is told in great detail in Valerie Barrow's *Alcheringa*, which is a planetary history book. It details the arrival of a huge Pleiadian mothership, the *Rexegena*, which was sent to Earth with 50,000 people on board consisting of a large crew and their families. The Lyran Commander-in-Chief Alchquarina, and his wife, Egarina, were to intervene in an ever-worsening situation involving the Draconians and their Repterran offspring who had hijacked planet Earth. (See page 113.) These beings had set themselves up very comfortably here, and were experimenting in genetics by creating more and more savage creatures, basically turning the planet into an inhospitable war zone. Such a situation could not be allowed to continue, so the *Rexegena* and her crew were sent to investigate and intervene.

The Reptoids found out about this plan and pretended to cooperate in order to lure the Pleiadians within firing range. Sonic weaponry was used to destroy the Rexegena in the same way that a high-pitched sound can smash a glass. The ship had been created by its builders out of a crystalline material. The sonic frequency that was used to destroy it had the effect of melting the ship, with portions of it falling to Earth and scattering in the area around Bohemia and Moravia, in what is now Czechoslovakia. Of the 50,000 personnel on board, a number were able to escape the initial attack in small "rescue pods." Most of these were also shot down, but one with 90 people on board managed to escape and land in the sea off the coast of what is now south-eastern Australia.

The melted remains of the mother ship fell to Earth, solidified, and became the strange translucent green Moldavite stones that geologists classify as meteorites, which is not the whole story. People whose souls were on board the Rexegena feel a deep emotional response when they touch Moldavite.

The small group of survivors encountered severe hardships. Many were badly injured, and they all had to contend with the heat, humidity, dangerous animal life, hostile Reptoids and the much heavier atmosphere of Earth. They struggled to breathe, and became exhausted with the least exertion.

They were stranded far from their home world, most of their survival and medical equipment had been lost, and they were faced with the prospect of dying before their mission could even be started, let alone accomplished. As many people do when faced with such a situation, they felt a need to record the events and circumstances of what had happened. That way, if anyone came looking for them in the future, there would at least be evidence and an explanation of their plight. Their written language was in the form of hieroglyphs, some of which still remain on cave walls in a place now known as Kariong, north of Sydney, New South Wales.

With the situation they found themselves in, there were two choices. First, in a bid to help the unfortunate "mixture" creatures, these surviving ETs decided to attempt further genetic manipulation in order to upgrade the hybrid beings into a more fully human form. In this way, their energy frequency could be shifted more completely from animal to human. Secondly, the resulting offspring would carry ET genetics, but they would have a better chance of surviving the challenging and harsh planetary conditions than the ET progenitors themselves.

They had very little equipment at their disposal, but they were experts at this type of work. By contributing their own genetic material, and with some of the females willing to use their own bodies as surrogates, several primate/reptoid/Pleiadian offspring were born. However, the Pleiadian surrogate mothers' bodies weren't compatible for carrying these mixed-genetic fetuses, which were born with massive problems. Those that survived the birth process died soon after. The only alternative was to capture some adult females of the "mixture" slave species. The upgraded embryos were placed into their bodies, which were closer in energetic frequency to the offspring the Pleiadians were trying to produce, which is the same reason that in the present-day ET hybrid program, Earth-human mothers are often brought in as surrogates.

The Pleiadian/Lyran/human offspring evolved into the first *Homo sapiens sapiens* born on Earth. Thus "Adam and Eve" were the forebears of the original people, now known as the Aborigines. Australia — believed by some scholars to be the oldest country on Earth — may even be the sought-after "Garden of Eden," where animal man became *hu*-man.

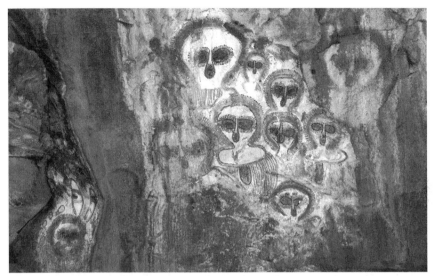
ABORIGINAL PETROGLYPH, ELIZABETH STATION, AUSTRALIA

Supportive of this theory, in King's Canyon near Uluru (Ayres Rock) in Central Australia is an area known as "The Garden of Eden." This place is highly unusual, as it forms a lush tropical oasis in an otherwise barren desert landscape. Tourists come from all over the world to visit this magical spot.

However, the initial development of the earlier proto-humans, the slave races, had begun millions of years earlier by the Draconians and their Repterran offspring in places such as Africa and the Middle East. In fact, Middle Eastern civilizations were influenced strongly by a Repterran culture that came to be known in the local language as "Anunnaki." These were a hybrid reptoid/bird humanoid group also known as "Nephilim" because of their "fallen" status. Consequently, very ancient part-human remains are found in these places. The famous "Lucy" that was found in Africa is an example of these very early "mixtures," so she could certainly be one of the first mammalian primate/humans developed by the Reptoids. However, it may be that the first modern-day human being, the "Eve" spoken of in Genesis, was really born in Australia. This upgraded *Homo sapiens sapiens* species, with its enhanced off-planet (Pleiadian as opposed to

Draconian reptoid) human genetic component, then eventually spread out across the rest of the planet.

Eventually, to assist in the work of upgrading the slave species to a higher vibrational frequency, Zeta Reticulans, Sirians, and other off-planet people joined the Pleiadians and Lyrans. Hence the cultural memories and records that have been passed down in various tribal systems around the world claim different Star Beings/Sky Fathers as their ancestors.

During the time of Repterran control and human slavery on Earth, the Ant People had been driven far underground and off-planet. They kept very much to themselves, hidden away in passages deep beneath the surface, having suffered so much at the hands of the Reptoids. The arrival of the Pleiadians and Lyrans gave them renewed hope. Some of them ventured out of hiding in order to help the new arrivals, many of whom were injured, unable to cope with the harsh environment and struggling to survive. The Ant People (early Greys) took the survivors down into their underground shelters to keep them safe from the Reptoids, and provided them with medical aid. They were knowledgeable in genetic work, ecology, and the adaptation of species to foreign planetary environments, so they were able to lend their assistance and equipment to the upgrading project.

The entire Australian Aboriginal "Dreamtime" history, as recorded by many diverse Aboriginal cultures in many different languages, is firmly based on this story of Ancestor/Creator Beings who came to Earth from off-planet. All Aboriginal people acknowledge these off-planet cultures as ancestors. The Rainbow Serpent is representative of the DNA and genetic work carried out by these highly honored Beings whose likeness is recorded in the sacred Australian Wandjina drawings.

"They say we have been here for 60,000 years, but it is much longer.... We have lived and kept the Earth as it was on the First Day. All other peoples in the world came from us."

— Australian Aboriginal Elder, quoted in Cosmic Cradle

Repterrans Persevere

The Repterrans, who had already developed the slave race and were continuing their residence on Earth, considered themselves to be the dominant human group and owners of the planet. As they witnessed this new and upgraded human species being developed they felt threatened. They wanted the newer human species to remain as disempowered slaves, so their plan was to block the process. Their act of interference is recalled in the Bible as the "temptation of Eve in the Garden of Eden."

The "Fruit of the Tree of Knowledge" that "Adam and Eve" partook of was the acquisition of reasoning power and free will, which was being introduced very gradually through DNA activation. This reasoning power, or "knowing," is part of the evolutionary path from the purely instinctual drive of the animal kingdom to the reasoning power of the human. However, with reasoning power comes the responsibility of free will. Since the two work together hand in hand, it is a vital key to evolution on a level-one planet such as Earth.

Because the New Humans were not quite ready to handle this responsibility, the Repterrans interfered by bestowing reasoning power and free will upon the newly evolved mammalian humans before they were able to handle it wisely. Thus instead of helping human evolution, the Repterrans hindered it, resulting in the "fall of man."

The "humanizing" process involves enhancement of the DNA. It is purely scientific, and is carried out through advanced genetic engineering. In this area of energy transformation, spirituality and science merge. The bestowal of reasoning power is certainly part of this assisted evolutionary process, but, in this case, it was introduced too early. Gregg Braden, in *Human by Design*, reports that science now accepts that human DNA was edited from chimpanzee DNA in two chromosomes, 2 and 8, some 200,000 years ago. One edit enabled complex spoken language, and the other enhanced human reasoning. Braden discusses how and why, but does not yet venture to ask, "Who were the editors?" That answer is still too out-of-the-box for most scientists to openly discuss, because the implications knock the legs out from under the old scientific paradigm.

The wrong choices made by the New Humans through this prematurely bestowed gift of free will are symbolized in the two "sons" of Adam and Eve—Cain and Abel. Cain is symbolic of the lower path of fear (jealousy), whereas Abel symbolizes the higher path of unconditional love. Cain overcame Abel and killed him, which is an allegory representing the lower path that was chosen by the New Humans.

The Repterrans, personified in the Bible as Lucifer, which means "Light Bringer," enhanced the development of the New Humans in one area but in another way blocked their spiritual growth. This is exactly what's happening now on Earth, with some humans being given psychic and clairvoyant powers or advanced technology by the Repterrans before they're spiritually ready to deal with such responsibilities wisely. This interference is being done by energy manipulation of the pineal gland via the "third-eye" chakra point in order to throw things out of balance, and so the cycle repeats.

DAUGHTERS OF MEN

The off-planet people who were working in a positive way developed the New Human species by artificial insemination. However, the Repterrans were caught up in the physical and emotional needs of their biological bodies. Therefore, they began physically mating with the newly developed mammalian humans. To put this in biblical terms, the "*fallen* Watchers" (Anunnaki/Nephilim) took the "daughters of men" to themselves as wives. The resulting offspring are also known as "Nephilim."

This term translates as "fallen" or "descended," so, in one sense, it refers to the fallen (devolved) Watchers, and, in another sense, it could refer to their offspring (descendents). It has also been translated as "giants," which is also correct, because they were, like most reptoid humans, much larger and taller than modern Earth humans. The giant skeletal remains of these and other beings have been found in many places on Earth.

Stories of gods and demi-gods mating with humans abound in Greek and Roman mythology as echoes of actual happenings. An example is Leda being mated by Zeus in the form of a swan. After all, the Anunnakis were half reptoid/half bird people. Quetzalcoatl, the "feathered serpent," is another example of such a godlike hybrid being.

By the time the new *Homo sapiens sapiens* species was developed (50,000 to 100,000 BCE), the Repterran humans had become an ancient race on Earth, having been around for millions of years. They had inherited a high level of technology from their Draconian forebears that included rocket science and space travel, which caused the New Humans to look upon them as "gods," travelling around in their "chariots of fire." Unfortunately, this worship was not so much because of the Repterran's spirituality, but rather because of their high technology. Arthur C. Clarke famously expressed it this way: "Any sufficiently advanced technology is indistinguishable from magic."

As for the Repterrans, many of them were so caught up in their own power and ego that this worship by the younger human species was not only encouraged but insisted upon. They had no interest in correcting the misunderstanding. In many cases the "God" of the Old Testament was actually Repterrans playing a role aimed at keeping the New Humans under control. Divide-and-conquer techniques were effectively designed to incite wars between various human tribes — as they still do today!

Sasquatch Elder Kamooh: "The Anunnaki have kept coming to Earth over the ages, even into human times, when they were often worshipped like gods, with the imposition of cults, through fear or force. They are those gods who demanded bloodshed and human sacrifice, required gold, taught humans to cast spells, and led masses of humans into wars. They later established their dynasties over your people, establishing their hybrid bloodlines as your rulers....

"Over eras, your populations and societies grew and diversified. In this process, diverse religious cults and groups were created, often influenced by the lower lords. They have devised or twisted many belief systems to control humans through your minds and emotions.

"As you still can see among your people today through religious extremists, this kind of indoctrination to control masses causes much division, conflicts, violence, pains and sufferings, at a time of your evolution when your collective consciousness is emerging from ages of sectarianism and religious racism."

The Repterran/Anunnaki also encouraged the New Humans to believe that they'd been "created in the image of God," which in reality was just the image of the Repterrans themselves. All human species throughout the universe have the same basic hominid body shape, although in some much older species like the Greys, who have almost evolved past the need for physical, biological form, the mouth, nose and ears are vestigial. The body is much simpler and the head is larger because of the higher level of mind/conscious awareness.

The Book of Enoch describes those fallen Watchers (Repterrans) as being tall, fair and handsome-looking people. The Repterrans are shape-shifters, so can take whatever form they want to, including an attractive human form. The myth that physical beauty denotes "good," and "ugly" (in human eyes) denotes "bad," has been promoted on Earth by the Repterran Controllers for many millennia. It drives the fashion, makeup and film industry to this day! The Repterrans continually take advantage of how Earth humans can be so easily manipulated by a being who looks "angelic." In Greek legend they're recalled as the Gods of Olympus, who are idealized as being perfect in human form, and with Godlike powers. They were also the basis of Hitler's "Aryan Ideal." The Repterrans were a major part of the Atlantean culture on which the Sumerian, Egyptian and other such cultures were based. The Sumerian word "Anunnaki" was the title given to them in their role as the founders of those cultures that were carried over from Atlantis. By that time, many Repterrans were reincarnating in Earth human form here, as they continue to do now.

By hijacking planet Earth many millennia ago, the Repterrans have caused major problems between themselves and the other off-world people. The more highly evolved interdimensionals and ETs are universal caretakers and custodians whose job is to oversee the development of life on many planets. The Repterran takeover of Planet Earth was an act of blatant rebellion against a universal code of conduct. This rebellion continued to escalate throughout the Atlantean era. The Repterrans' genetic experimentation and development of all sorts of hybrid creatures was totally unnatural and strictly against universal law. Even today Repterrans continue to reincarnate on Earth in Earth-human form and refuse to relinquish their ownership and complete control of Earth and humanity. For this reason many refer to them as the Controllers.

Many of the Repterran Controller group and their descendants survived the demise of Atlantis, which is believed to have occurred around 10,000 to 12,000 BCE. They did this by escaping to bases they had previously established both inside Earth and on Mars. One of the mountains on Mars, Olympus Mons, was named after Mt. Olympus, which is said to have been the home of the Gods of Olympus (Repterrans). Because the Repterran survivors of the Atlantean cataclysm (Biblical Flood) had a strong influence on Egyptian and other cultures in that area, the pyramids and other structures on Mars correspond to similar structures on Earth.

Because the Repterrans are patriarchal, their influence caused many cultures and religions that began in the Middle East to also be highly patriarchal. Other more spiritually advanced ET cultures like the Zetas, Pleiadians and Arcturians are totally egalitarian, with both males and females regarded as being equal on all levels without discrimination.

The so-called Inner Earth is not a physical place — it is interdimensional with gateways to the astral realm. Some of the Repterrans took advantage of this as an escape route, while others have continued to reincarnate on Earth as Earth Humans. However, there is a vast network of tunnels, as well as large subterranean caverns and constructed underground bases.

Having claimed Planet Earth and the New Human race as their own, Repterrans have used weapons to block any attempts by the off-planet groups to return. Large areas of sand that have been subjected to enormous heat have been found in areas of the Middle East and Northern Africa, as well as in the Gobi Desert region. These are the results of an off-planet war that took place about 8,000 BCE.

This conflict was in connection with an Inner Earth base created by the Zetas and Pleiadians who came to Earth to try to get Earth out of the Controllers' hands. This base came under a surprise attack and was destroyed. The legend of "Shambhala" still recalls this magical place in the mountains and under the Earth. It was inhabited by advanced off-planet teachers who had come to Earth to introduce the Vedic Teachings and free people from the Controllers' power. These Vedic Teachings are the basis of early Hinduism, from which Buddhism developed. Of all Earth-plane religions, Buddhism comes closest to higher off-planet spiritual systems.

The Guardian crew staffing the Shambhala base in Northern India suffered a similar fate to many of the survivors of the Rexegina. We need to bear in mind that, as spiritually advanced as they were, they were still physical human beings carrying out a dangerous mission on a hostile planet. Even Jesus, a very highly evolved Star Being, was put to death down here. Star people and their ships are still being shot down and imprisoned by the Repterran-controlled authorities here on Earth. This is the reason why more and more off-planet humans are now "coming in the back door" by volunteering to be born here in Earth human form.

The Shambhala crew were either put to death or their spiritual essences were forcibly removed from their Grey containers and placed into Earth-human ones. They were then driven from the base, and the survivors fled westwards carrying their esoteric knowledge and sciences to ancient Egypt. Here they were honored as priests and priestesses, until the Controllers took over and again attempted to destroy them along with their teachings.[1]

---

1. Author's note: Proto-Egyptian hieroglyphs were based on the written form of the Pleiadian language, as evidenced by the hieroglyphs and symbols in cave paintings and on remains of recovered starcraft.

The Egyptian story of Osiris, Isis and Set portrays how the Guardian culture was overtaken by the Repterran Controllers. However, this story is actually based on an even more ancient account of an attack on the Zeta Reticulan culture by the Reptilians and their Blond allies (a rebel Pleiadian group).

As the legend is recounted on Earth, Osiris was set upon by the evil Set, who is depicted as a reptile (crocodile). Set is said to have chopped Osiris's body into pieces and thrown them into the Nile. Isis, his twin sister and wife of Osiris, managed to retrieve the parts and put the body back together again, all except for the penis, which couldn't be found. The missing penis is symbolic of the disempowerment of the Reticulan culture and their inability to reproduce. Osiris being cut to pieces is symbolic of the Reticulan cultures that were torn asunder by the Reptilians and Blonds. They've mostly managed to survive and recover from this attack, except for their ability to reproduce naturally. A type of ethnic-cleansing program was carried out on them, which destroyed their reproductive organs.

Once more they were forced to flee to the west. These off-planet refugees were the original Romanies[1], also known as "Gypsies," which is a shortening of the word "Egyptians." Despite Controller infiltration into their ranks and much interference over time, a few of these "Gypsy" descendants have still managed to carry the teachings down through generations, along with the understandings of their true origins.

A Romany legend tells of a group of these "Gypsy" descendants, led by Sara, their tribe's matriarch, who travelled to the Camargue region in Southern France in the year 1 AD. They made the journey to meet boats carrying Mary the Mother of Jesus, Mary Jacobe and Mary Magdalene. Saint James, the brother of Jesus, along with John the Divine, Joseph of Arimathea, and other members of Jesus's family, also arrived on the boats, fleeing for their lives after the crucifix-

---

1. See Peter Moon: *Transylvanian Sunrise* and *Transylvanian Moonrise* and the Romanian Sphinx photo on the cover!

ion of Jesus. This location is now known as Les Saintes-Maries-de-la-Mer, honoring the three Saint Marys.

The Gypsy Guardians took the family into their tribe and provided them with guidance and protection. This meeting of the nine boats carrying the family of Jesus is celebrated and re-enacted every year on the 25th of May by the Romany people. Travelers from all over the planet gather there to commemorate the escape to safety.

A number of attempts have been made to get Earth out of Controller hands, but they are deeply entrenched here, and they have the younger human species so completely under their control that all past attempts have failed. The Pharaohs Akhenaton and Tutankhamen were off-planet people who came down here in hybrid containers chosen specifically at that time to enable them to remain for the span of a lifetime in human form. Although they tried to set things right, they were both put to death. These weren't the only Guardians who have attempted to reverse Reptilian control. Others in the form of various spiritual teachers such as Jesus, Buddha and Krishna have tried as well.

Another intervention attempt occurred when a mysterious image of a white horse led a priest named Gantei to climb Mt. Kurama in Japan. This off-planet visitation, which was a classic ET encounter, took place in the year 770 AD and was the first step that ultimately led to the discovery of Reiki.

As a result of this encounter, Gantei founded a temple on Mt. Kurama in which there is a shrine dedicated to the Trinity known as Sonten, said to be the source of all creation and the essence of All That Is. According to legend, Sonten came to Earth over 6 million years ago in the form of a being known as Mao-son, who descended upon the mountain from Venus. His mission was the salvation and evolution of humankind and all things living on Earth.

For this reason, the founder of Reiki, Mikao Usui, came to Mt. Kurama to meditate and receive the necessary initiation to the Reiki Master symbol that is the key to accessing Reiki. This symbol is formed by the Japanese kanji symbol for Sonten. As with Gantei, Mikao Usui received this initiation and symbol via an ET contact experience.

UFO SAND SCULPTURE AT MT. KURAMA

Even though none of these attempts at saving humans succeeded, each one has added to the other to keep influencing current Earth humans to keep evolving.

THE ASTRAL PLANE

The Repterran Controllers have a strong presence and influence on the astral plane, which was set up originally by the assistant creators as a holding ground for Earth-Human spirits between physical incarnations. The astral plane is known as the Plane of Dreams, Emotion and Illusion. It is actually quite chaotic because of the fear and superstition that has been created in human minds by the Repterran Controllers. They've also interfered with the Akashic Records of Earth in order to create and maintain certain beliefs in the collective human consciousness that support the Reptilian agenda. Attachment to these Repterran Controller belief systems is a major part of the Earth-human psyche. A typical example of their interference in this area is some of the information that's been accessed in connection with Atlantis. Despite claims made by psychics, the Atlantean age was not a particularly spiritual time period or "Golden Age" on Earth. Its latter times in particular were decidedly unspiritual, as were some of the Egyptian belief systems that were totally astral plane/Controller-based.

Sasquatch Elder Kamooh: "It started back in the days of the downfall of Atlantis, when the lower lords, with the help of the Lizard-People, started cloning their own DNA to allow more souls of the lower realms under their control, to enter our physical plane."

Ancient Egypt was not completely a time of high spiritual understanding. An example was the mummification process, which was really aimed at keeping humans trapped in physical form, by placing great emphasis on preserving the physical body. Material possessions were carefully buried with the body so as to draw the entrapped spirit back for yet another life.

The Egyptian Mystery Schools were originally set up by the off-planet Guardians who had escaped from their Base in Northern India. The Guardians set up these centers of learning to protect ancient knowledge from Controller influence. However, many were infiltrated over time, becoming increasingly Controller-based. Many of their teachings involved the manipulation of astral plane energy, which enables one to perform seemingly amazing magical and miraculous feats. In many cases, the highly esoteric "Mystery Teachings" that can be revealed to only a select few are simply teachings whose origins are off-planet and therefore perceived as being beyond mortal abilities. The positive aspect of these Mystery Schools has been the preservation and protection of true teachings such as the reincarnation of the soul. It is unfortunate that the concept of reincarnation was later removed from Christian doctrine when the Church was taken over by the Controller Emperor Justinian around 550 A.D. to form the Holy Roman Empire.

These Controller Emperors were responsible for removing and altering much of Jesus's teachings. They created a religion based on their own agenda of patriarchal domination and control of the masses through fear and superstition. The Church and other governing systems through the Dark Ages were totally in the hands of the Controllers, and the Controllers removed many truths from most mainstream religions on Earth in order to stop the followers of these religions from accessing the necessary depth of understanding and spiritual enlightenment to get themselves past the illusion traps of the astral plane. Too often these teachings have been replaced by fear, ego-oriented dogma and distorted versions of truth that cause people to experience pre-programmed illusions during the time they spend in the astral plane between lives. Their memories are then erased so they have no concept of the time between lives or past incarnations. They are then "recycled" back to Earth to again become enmeshed in materialism, neediness, fear, ego, superstition, *etc.* in yet another round of physical existence, unable to evolve to higher levels.

Sasquatch Elder Kamooh: "The souls instructed this way (Church doctrine denying reincarnation) lack the necessary understanding and motivation to continue consciously their spiritual evolution on this physical plane, as well as in higher-dimensional planes in the afterlife. The long-term purpose of incarnations is the evolution of the individual soul and ultimately, of the Greater Soul and Universal Consciousness."

Some positive intervention happened at times such as the Reformation and Renaissance, when more highly evolved beings volunteered to be born on Earth to try to right the wrongs and introduce more enlightenment. The Controllers however were too deeply entrenched, and generally either imprisoned or killed these beings, as happened with Jesus, Galileo, Copernicus, Giordano Bruno and many others, as well as modern-day reformers like Martin Luther King, Jr., Mahatma Gandhi, and Nelson Mandala. Despite any persecution in their lives, these enlightened reformers (and others) created major change for the human race. The downside is that the Controllers then take control of the new and improved systems that the newcomers have set up, tailoring them to once again exert maximum control and exploitation of people. But it seems to be getting harder for them as we evolve.

Despite the control of religions on Earth, there are still millions of people who follow the common tenets that flow through them all, mainly that love is the overwhelming and most powerful energy on Earth.

If you are feeling a little bleak right now, hang on and read on. Later on in the book you'll learn that the biggest problem is fear and the Selfless ETs will show us how to evolve past that and acquire personal control. They have been trying to teach us this for millennium, and they are not giving up. Even though it didn't work before, it can now as we are more highly advanced spiritually than ever before.

9

The Other Players

Many on and off-planet humans have been involved with Earth and I'll present the major ones here. I know a lot about the Arcturians and Zeta-Reticulans, however, Sanni Ceto is much more consciously knowledgeable of all the ET races than I am, so I'll borrow from her to give a brief description of the other groups who work closely with the Zetas.

ARCTURIANS

The blue-grey Arcturians are the ones I'm most familiar with, but I understand that there is also another more Earth human-looking group with a bluish skin color. Again, several different planets and races are involved. The blue-greys' height is around five feet for males and four feet for females. Like all Greys, there isn't as great a difference between genders as there is on Earth. They're highly telepathic, and they work as emissaries to other planets and galactic groups. They also carry out exploration work dealing with environmental issues on many planets.

The Arcturians life span is around 600 Earth years, and their home is a medium-sized planet outside of the Arcturus star system. Several colonies also live in the Rigel Kantares system, as well as a few who live in the Betelgeuse system.

The home planet that both Jacquelin (Zan Tu Kai) and I (El Or Kah) can recall is a tropical world with vegetation such as palm trees and bamboo, which was, and still is, widely used as a building material. Both of us in our present lives had a childhood obsession with drawing palm trees, and we like to use bamboo furnishings in our homes.

LYRANS

Lyrans are a race closely related to the Pleiadians and Earth humans. They are brown- or gray-eyed and tan colored, with reddish-brown hair. They are also the record keepers who maintain large storehouses of information. They were one of the seed colonies sent to terraform Earth after the great meteor impacts that wiped out so many species in prehistoric times. They are scientists who conducted genetic programs to create many plant and animal species on Earth and other worlds.

Like the Meropian Pleiadians, they are cousins to Earth humans and reproduce as Earth humans do, with a gestation period of about 12 months. They are very fond of music and the arts, and their temples, which hold the planetary records, are like crystal pyramids in appearance. They eat foods similar to what Earth humans eat, but they are totally vegetarian.

PLEIADIANS

Since Earth-humans and Pleiadians are two very closely related races, Earth humans generally feel more comfortable with the Pleiadians than they do with Zetas and other Greys. Pleiadians are very comfortable with Zetas and other Greys, as they all work closely together. The Pleiadians are like cousins to Earth humans, who share much of their genetic material.

There are several different groups of Pleiadians, but those most closely associated with Earth visitation are the Meropians. Their home world, Merope, is a forested planet with abundant water and a climate similar to that of Earth but with less severe winters.

Meropians are taller than Earth humans, with the males standing about seven to eight feet and females about six feet in height. Their eyes are clear blue, and their skin is a pale, milky white. Males have golden-blond hair and females have a more platinum color. Their children's hair is white, which darkens as they mature.

They live mostly in villages, and their system of government is similar to Earth's, but there are no wars. They are totally peace loving, and they have selectively bred out undesirable traits like aggression. A strict social order is maintained for work and food cultivation, and the children are schooled by their parents under the guidance of androids programmed by the local government.

Meropians eat meat and also consume vegetables and fruits. Many of these plants have been introduced to Earth. The main source of protein is a species of bio-engineered fish, and the people live in ultra-modern homes where everything is fully automated.

Pleiadians and other humanoid ETs make use of Grey soul containers when they need to carry out work on planets that may not be environmentally suitable for them. For example, in *Alcheringa*, Valerie Barrow relates how a Pleiadian and Lyran group that became stranded in Australia 900,000 years ago were using Grey containers as "space suits," hence the Wandjina drawings in Aboriginal art that depict the Creator/Ancestor Beings as looking like Greys.

REPTILIANS

Like the Greys, there are Reptilian cultures on a number of different planets at various levels of spiritual evolution on the *Human Ladder*. There are some not-so-pleasant Reptilians on other Level-One planets (like Planet Earth) that have caused the Zetas problems in the past because, even at level one, the Zetas had mastered space travel outside of their own "safe zone." This "safe zone" refers to the vibrational frequency around an inhabited planet that stops less evolved groups from attacking the planet's inhabitants. Any beings who space travel beyond their planet's protective frequency zone may leave themselves open for attack or interference. Suffice it to say that if Earth humans conduct deep-space exploration before they evolve further up the Human Ladder, their efforts may fall victim to interference from less-evolved beings.

Other Reptilian cultures are highly spiritual beings that exist on levels two, three and upwards on the Human Ladder. They should not be confused with the Repterrans (a reptoid human species developed on Earth by the Draconians). Some of these higher Reptilian cultures are assisting Earth with the shift upwards to become a Level-Two planet.

According to Sanni Ceto, "They are generally scaled, with vertically slitted pupils similar to cats. Two of the more evolved reptoid cultures are the Ter-hig-gom or Tar-hig-gom, and the Driseih.

"The Tar-hig-gom are large people, the males standing about eight feet tall and the females about six feet. They trade with the Zeta colonies and act as peacekeepers on various worlds. Most are telepathic, but they also use vocal speech, which is like a high-pitched whine. Their life-span is about 890 Earth years for men and about 899 Earth years for women.

"The Driseih stand about 7 to 8 feet tall, and their home world is a large planet that orbits a single star in the Antares system. They live in large communities ruled by a head or ancient elder who is like a king on Earth. Theirs is a patriarchal system. They are part of the Planetary Alliance and take part in trading with other cultures."

ZETAS AND GREYS

Some researchers on Earth think of the Greys and Zetas as being all one race, which is not correct. The Zeta Reticulan culture is actually composed of a number of different planetary groups who reside in the Zeta Reticuli star system, as well as Zeta-type beings living on other planets outside of that system. Even within the Reticulan system, there are around 30 different cultures, and some are more highly evolved than others.

The Greys are an insectoid-human species, some of whom first evolved on Planet Earth *billions* of years ago, *viz.*, the Ant People, referred to by Sasquatch Elder, Kamooh. The Ant People are also mentioned in the ancient Indian text known as the Mahabharata. Highly evolved off-planet insectoid assistant creators introduced DNA into them, resulting in rather different-looking human forms from the mammalian-based human species developed eons later on Earth.

There are mammalian/insectoid hybrid groups living in the Zeta system such as the Essassanis, who are of mixed Zeta and Earth-human genetics. The pure-blood Zetas, however, are one of several cultures referred to by Earth humans as Greys. The Zeta Reticulan Greys are known as Kebbans, who, like the Mantis people, are also a highly evolved, gentle and peace-loving society.

Insectoid humans generally use telepathy among themselves, and also in communicating with other species, but they cannot talk like Earth humans do, because their vocal cords are rudimentary and have been adapted out for the most part because it is not needed with telepathy. Some still do make chirping or whistling sounds similar to crickets or the clicking sound of dolphins, to which some Greys are related. Earth humans often can't pick up on telepathic communication, especially if they're in a state of fear, so the Zetas and Greys have great difficulty communicating with them during contact experiences.

As Sanni Ceto explains in *Zeti Child*, all the insectoid people have well-developed brains and nervous systems and are not as large in stature as most humanoids and reptoids, having lightweight bodies designed for intergalactic space travel so the G forces won't harm them. However, Oris, the Elder featured in *The Zeta Message*, is nearly 8 feet in height, as is my own teacher, Maris, but these beings generally travel out-of-body. The Mantis people in particular are very highly evolved human-type beings who can be very tall, but often travel out of body. When they are seen, many people have reported Mantis beings to be seven or eight feet tall.

Most Earthlings can trace their family lines back to several different nationalities. So it is with off-planet people, of whom most are hybrid mixtures of at least two different planetary species. When Earthlings' awareness increases about their off-planet ancestors, they will also add these different planetary species to their ancestry.

Sasquatch

According to SunBôw, author of *The Sasquatch Message to Humanity*, these beings are also known as Bigfoot, Yeti, Almas, Yowie and

Skunk Ape, as well as other names. An association can be made with Hanuman and the Ape People of the Vedic scriptures in India or similar stories transmitted for centuries. They are emissaries between our world and the underworld (Inner Earth dimensions).

Sasquatch Elder Kamooh: "When DNA star seeds from our Star Elders from the Pleiades, as well as from the Great Bear [Ursa Major], where a similar species to ours had been living (the Tageui, according to Sanni Ceto — Judy), added to our DNA the unknown gene that your scientists can't understand, my Sasquatch People were conceived in the underground temples of Agartha. You know us as Sasquatch, from the language of your Salish tribe who have continued knowing us, but my Hairy People were known by many other names then, as it has been since. We are the Mammal People.

"Our spiritual mission has been to be protectors and caretakers of all life forms, and of the spiritual knowledge of our Star Elders, by saving in our memory their teachings of the long Soul evolution. Of all hybrids created on Earth, only my people, some of the Bird-People in their guardian-angel forms, and the Elementals (including Greys) have remained faithful to our original spiritual mission, in spite of mistakes along the road. This is why we still gather with the Elementals and Little People to celebrate our spiritual bonds, and we wish for your Human people to join with us and reintegrate the Council of Star Elders with us."

The following beings form a class all by themselves, because their role on Earth is very different from all the others:

THE CONTROLLERS

The Controllers, or Repterrans, which is another name I use for them, are an Earth-human group descended from a technically advanced off-planet group that settled on Earth millions of years ago, long before Earth humans were here. These original ones were the Draconians, a reptoid human species native to the Dorado and Pictor sectors, with some colonies scattered in the Orion Nebular sector. They were responsible for seeding the planet with the dinosaur species, then they went on to bio-engineer a number of biped human/reptilian beings from dinosaur stock, three of which are the De-raks, the Te-raks and the Puy-yats. I refer to these ones as Repterrans, because they are a much older reptilian human species developed on Earth (Terra, or Terai in the Zeta Reticulan language) as opposed to modern-day Earth humans, who were developed later out of mammalian (primate) stock.

Sanni Ceto describes these Earth-plane-developed reptoids this way: "De-raks come in several colors and are reptilian-type people with males having a crest on their heads and females being smaller and lacking a crest. They are shape-shifters and pass themselves off as modern Earth humans as a way of controlling the authorities down here.

"Te-raks are mostly subterranean dwellers that resemble a monitor lizard/human crossbreed. The males are eight to nine feet tall and walk upright on two legs. They speak by telepathy, and like the De-raks, are highly aggressive. They are also shape-shifters, and work with the military in secret underground tunnels and bases on Earth.

"The Puy-yats have large red eyes with black slit pupils. The face is a deep cherry to chestnut brown in color. The nose and forehead are dark green with lime-colored cheeks, pockmarked in males and smooth in females. Their height is about seven to eight feet for males, and five to six feet for females. They are large, slender reptoids related to the De-raks and are descended from a dinosaurian-type reptilian that was native to Earth. They were removed off the planet by the De-raks before the great extinction that caused the demise of the ruling reptiles on Earth. A brain parasite deliberately created by the Te-raks was at least partly responsible for this extinction, as well as the fallout from a giant meteor strike which affected the world's climates by generating major volcanic activity. The ash cooled the planetary temperatures by blanketing the Earth and blocking out the sun's rays. The meteor strike also caused a tilt in the axis of the planet. This in turn resulted in a shifting of the winds and currents.

"When the original Draconians first came to Earth millions of years ago and established themselves here, they were a more highly evolved species, but they too became infested with this parasite, which affected them by feeding into the negative emotion centers of their brains, causing them to become hostile and controlling, thriving on war and aggressive physicality.

"When this occurred, they devolved to a lower-energy frequency, thus becoming a "fallen race." These Draconians interacted with early Earth humans, which gave rise to the legends of Satan and his people. The parasite was also used against the later mammal-based human species, and it affected them by shutting down a part of their developing DNA, thus effectively blocking the humanizing process that other assistant creators were attempting to carry out.

"The Draconians and some of their offspring became trapped down here, and because they're still caught up in fear, ego and a need to control, they haven't been able to move on and evolve. They just keep getting pulled back down here on an endless cycle of reincarnation, generally into Earth human form. In this way, they've created a vicious circle for themselves that they can't break out of, and have become completely caught up in their own fear, and the need to keep others down in fear along with them. Earth is one of the lowest-frequency planets, and the fear is feeding upon itself."

A good analogy used by the Grey teachers is that the universe is like a river that's flowing along. However, the Repterrans hijacked the Earth many millennia ago. Consequently, Earth humans have been pushed off to the side as in a stagnant pool, cut off from the rest of the river. This has stopped humanity from spiritually evolving as quickly as was intended. Thus Earth-plane humanity is out of sync with the rest of the universe.

The information on brain-destroying parasites was given to me by Sanni Ceto, but "coincidence" guided me to read Dr. Eben Alexander's, *Proof of Heaven: A Neurosurgeon's Journey into the Afterlife*, in which he describes the bacteria *Escherichia coli*, better known as *E. coli*. He says that no one knows how old *E. coli* is, but he estimates that it is perhaps three or four billion years old. He goes on to explain that the organism has no nucleus and reproduces by the primitive but extremely efficient process known as asexual binary fission, which simply means splitting in two.

Dr. Alexander says that *E. coli* is basically a cell filled with DNA that has the ability to take in nutrients directly through its cellular walls, usually from other cells that it attacks and absorbs. It can simultaneously copy several strands of DNA and split into two daughter cells every 20 minutes or so. In cases of bacterial meningitis, he reports, the bacteria attack the outer layer of the brain, or cortex, first, which in turn surrounds the more primitive (reptilian) sections of the brain. The cortex is responsible for memory, language, emotion, visual and auditory awareness, and logic, so, when an organism like *E. coli* attacks the brain, the initial damage is to the areas that perform the functions most crucial to maintaining our human qualities.

In other words, if you want to de-humanize a developing and evolving human species, just expose them to *E. coli!*

Physical existence only manifests on the lower rungs of the Human Ladder, because gross physical form has a lower-vibrational frequency. Even by the time we've evolved our spiritual frequencies to level three of the Ladder, we're beginning to move beyond the need for physicality. Most intelligent beings in the universe vibrate at higher non-physical frequencies, but they can still use physical soul containers when they need to. Earth Humans, on the other hand, are generally stuck in physical soul containers, believing that they are their bodies.

Remember anybody, even the most highly evolved beings, can potentially "fall down the Ladder" if they become too caught up in desire and physicality, which is exactly what happened with the Draconians. These are referred to in the Bible as the "fallen angels" particularly the one who was known as "Lucifer" who had "sat on the right hand of God." The original Draconian reptoids were quite highly evolved beings, until the fall. In other words, it can potentially happen to anyone!

The Repterrans' base is the Inner Earth, with its portals into the astral plane, but being energetically "stuck" here, the majority of them are incarnating physically as Earth humans. As one of the original humanoid species on Earth, they consider themselves to be the owners of the planet and superior to the more modern Earth humans, whose genesis occurred at a later time.

Kim Carlsberg in *The Art of Close Encounters*, p. 32, writes, "I have received at least 20 different descriptions of the reptilians, with just as many diverse elaborations about their demeanors and the types of relationships the experiencers have with them. The amount of information on them is quite extensive. Repeated over and over again, is that they are not from outer space, but rather ... they are from Inner Earth."

WHAT IS A HUMAN BEING?

Because humans have been developed from many different forms, the term human being has different definitions on different planets. The bottom line is that humans are an animal species that uses reasoning power as opposed to the purely instinctual drive of the animal kingdom. We have to be careful here, because the human ego can get pumped up if we suggest that humans can consciously recognize God Consciousness, while animals cannot. Other animals recognize God Consciousness, too, but in a different way. Many animals are actually more spiritually evolved than some humans.

"Man cannot assign a surviving soul to himself and deny it to his animal brothers; both man and animal are creatures of instinct and reason with the difference one of degree and not of kind; if consciousness does survive, it is quality of life and not of *Homo sapiens*."[1]

Humanoids always walk upright on two legs with one head and two arms. Some are fairly hairless, but then there are the Sasquatch People, who can be covered in hair, but are highly evolved human

---

1. Gaddis-Cowles, 1970.

beings as well. It's like trying to differentiate between a "horse" and a "donkey." We're all part of the animal kingdom, and there are physical differences, but the borderlines can sometimes be hard to reliably define.

According to author and Sasquatch expert, Kewaunee Lapseritis, there are seven races of this giant humanoid around the Earth. They preceded *Homo sapiens* by 50 million years, and part of their DNA was used to construct modern man. They are more like big brothers or uncles to Earth humans, rather than direct ancestors. The subterranean "lower lords" (Controllers/Repterrans) tried to exploit them as slaves, so they were removed from the 3-D Earth plane when the assistant creators came to intervene. The Repterrans seek to exterminate them, because having been helped to a higher evolutionary level by the ETs, they cannot be controlled as are Earth humans. They can live thousands of times longer than Earth people, they are highly evolved psychically, and they can easily move between dimensions, hence their apparent invisibility and the difficulty investigators have in finding physical proof of their reality.

As for the indefinite borderline between "human" and "animal," one can see this blurring of the boundaries between mineral, vegetable and animal kingdoms, for example, with such things as crystals, corals and sponges. Some forms of plant life like the Venus Fly Trap have taken on animal-like qualities, and some animals have taken on amazingly human-like qualities. Very few humans can practice the incredible depth of unconditional love as a faithful dog shows!

Fred Alan Wolf discusses the similarity between chlorophyll and blood in his book, *The Body Quantum*. "There is little difference between the structure of chlorophyll and blood hemoglobin. Apart from the core nucleus of chlorophyll being magnesium and the hemoglobin being iron, they appear identical." These comments suggest that both plants and animals may have a common ancestor.

All species have the potential to experience lifetimes in human form somewhere in the vastness of the universe. Here on Earth the human evolutionary process has come about through several species, including reptilians and mammalian primates. In other parts of the universe, other species have chosen to experience life as humans. Human beings are always a bio-engineered species, as is recorded in the Bible's Genesis story, but it needs to be understood that the biblical Genesis is an allegory that should not be taken literally. Most often outside assistance is required, as has happened on Earth.

10

Repterran Controllers in Modern Society

The Repterran Controllers are the ones known as the Elites,[1] or Illuminati, who have set themselves up in positions of wealth and power, keeping other humans enslaved through materialism, poverty, superstition and fear, which is often spread through religions based on truth but distorted to ensure that the Controllers remain in control. They're firmly behind the fundamentalist religions that deny the existence of any off-planet humans and label *all* ET contact as "demonic." They also operate through political, government and military systems employing Earth humans, but backstage-managed by Controllers.

As discussed in the chapter, The Players (see page 61), these Controllers were originally developed as humans on Earth during the time of the dinosaurs, so their form was reptilian-based, as opposed to the primate-based form of modern man, who was the product of a later process. You'll also recall that the Controllers are now incarnating on Earth in Earth-human form, spreading disinformation (fake news) of negative, off-planet reptilian-type ETs as well as Greys invading Earth, but, in fact, it is they themselves who are the ones carrying out invasive procedures and abductions on people. They have lived on Earth for much longer than modern Earth humans have, so they cannot be classified as *extra*terrestrials, *i.e.*, off-planet people. They *are* Earthlings. Many researchers believe the Controllers are responsible for the interference.

Admittedly they were members of the original off-planet Creator/Elohim Race, but are a break-away renegade group which has

---

1. Not to be confused with today's progressive thinkers who have been given this term.

devolved. Other Reptilian ETs have evolved to higher planes of reality and should not be confused with the Controllers. This disinformation campaign is taking place so as to enable the Controllers to remain as the dominant and ruling group on the planet, and to enable their continued exploitation and use of the younger human race for their own benefit. Consequently, they don't want humanity to evolve to higher levels.

The Controllers' whole aim is to block human evolution, and a major part of their agenda is to demonize the Greys, as well as the more evolved Reptoids and other ETs who are here on Earth at this time to help free humanity from the Repterrans' control and to assist humans up to the next level of the Human Ladder. Disinformation being put out about the Greys is that a) they are demons, b) they are just a mutated human race, and c) that they're hybrids created on Earth by the "Anunnaki," (the Repterrans who were the founders of the Sumerian and Babylonian cultures).

As explained previously, there are many varied human species throughout the universe, and not all have been developed out of primates. The Zeta Reticulan culture is a fully human species going back *billions* of years which did *not* evolve out of mammalian primate stock as did Earth humans. It was the Repterran Controllers who twisted the basic truth that all sentient life in the universe is "created in the image of God" by re-interpreting this concept to a more physically oriented idea of Earth humans *alone* being created to literally look like God.

I'm aware that I'm making some quite controversial claims here — claims that some readers may find confronting and hard to accept. For this reason, before going any further, I'll include some references from Paul Hellyer. I read Paul's *Light At the End of the Tunnel* not long after writing this chapter, and it provided some powerful validation on what I've written. With this in mind, I contacted him for permission to reference his book, which he kindly gave me. I respect what he has to say, as an ex-Defense Minister of Canada, on the subject of ETs and government. What he says also closely parallels my own understanding on this subject.

"There are several excuses given for maintaining secrecy about the ET presence and technology. These include religious concerns. Dr. (Michael) Wolf said that the Vatican had requested more time to condition Catholics and to say that man was not created to look like God, but that it is our souls that are created in the image of God." (p. 244)

Another excuse is that people might panic if they knew the truth. A final excuse is that switching too quickly from an oil to a clean-fuel economy would cause a great economic disruption and might cause the system to collapse, so the U.S. government feels it needs to maintain its near monopoly on some of the sophisticated ET-derived technology.

Please note: these excuses are not from Paul Hellyer himself. His comment is that none of these is a sufficient reason to withhold the truth. The aggressive and warlike tendencies of some Earth humans are very rare in other planetary cultures, for which the Controllers can be squarely blamed. Left alone or given the right teachings, Earth humans generally tend towards being peaceful, but because of the divide-and-conquer techniques and exploitation of their lower emotions by the Controllers, many are not given the chance to grow spiritually in the right direction. Some Earth humans do not even acknowledge the fact that they are far more than just a physical being and in fact are immortal souls who have chosen to experience physicality. This is a very sad state of affairs, which is why the Grey Guardians and many other ET and interdimensional beings are here now, trying to intervene when invited to do so. It is no accident that modern-day UFOlogy began with the appearance of ET ships not long after the detonation of the first nuclear device by modern man. The off-planet helpers do not want Earth humans to use nuclear weapons because they affect inhabited dimensions beyond human perception.

According to Sasquatch Elder Kamooh, "As we are all connected, the evolution of a species influences all the others on any home-planet and in the Universe at large. This is why the Star Elders are as concerned as we are about this Earth. Your human people, having lost a large part of their soul connection and spiritual understanding, are influencing in catastrophic ways every other life form on this home-planet and intelligent species in the Universe."

The Repterran Controllers still have bases in the Inner Earth, while infiltrating many levels of modern human society. They also have a strong influence on the astral plane. They still use divide-and-conquer techniques to keep Earth governments and people vulnerable, and they were the driving force behind the Nazi Regime, promoting the concept of the "Aryan Ideal." Like the "Gods of Olympus," the Controllers are still very focused on perfection of the physical Earth-human form rather than spiritual perfection or the acknowledgement of other advanced non-primate human-type forms throughout the universe. They practice remote mind control and therefore can affect the astral plane, including the Akashic Records of Earth, as this is a blueprint of the group consciousness and history of humanity, which has been manipulated and brainwashed by them for millennia.

Another example of Controller disinformation and fear- mongering is the information that has been spread around about Planet X, or Nibiru, which is said to be the home world of the Anunnaki, returning to our solar system to impact negatively on Earth. I believe some psychics caused panic by predicting its return in the first decade after the year 2000. The asteroid belt in our solar system, located between Mars and Jupiter is also spoken of in connection with the Anunnaki, as is the planet known as Maldek, or Apollyon. The approaching "body" that has been detected by psychics is not Planet X, but rather the Photon Belt, through which Earth is passing.

SunBôw, in *The Sasquatch Message to Humanity*, explains the real story behind these various predictions. According to Sasquatch Elder Kamooh, this event occurred long, long ago, during a time when an intergalactic civilization inhabited Earth. The Lizard people and the Bird People were caretakers of the highest civilization ever on this planet. The Ant People, almost wiped out after many wars with the lower lords who inhabited the underworld, had created a safe and peaceful off-planet home for themselves, which is now our moon. Spiritual centers visited by Star Elders had been built on Earth, and were used for education and healing, and also for the maintenance of Cosmic Order through the development of higher consciousness.

The majority of the lower lords had left Earth and settled on Mars and the next planet, known as Malkut, Maldek or Apollyon, as well as other names. They built star gates to enable them to import souls from other star systems they controlled. In response, the Star Elders created outposts on Mars, but the lower lords (Archons) remained firmly entrenched on Apollyon, where they continued to build their strength.

After many years of training and preparation on Apollyon, the Archons launched a massive invading army to Earth. At its head was a gigantic artificial "moon" serving as their mothership. The Star Council prepared to defend Earth. The first attack by the Archons involved a number of nuclear strikes against the Ant People's moon home. However, because of a previous destruction of an earlier moon home, the Ant People had reinforced the new one to enable greater protection, at least until the Star Council fleet could come to their rescue. Traces of those nuclear strikes and the craters thus formed can still be seen on the moon today.

The mothership/moon of the Archons was larger than our Moon and was constructed of metal. It came close to Earth, appearing huge in the sky and blocking out much of the sunlight. It terrorized the Earth population with loud sounds and powerful magnetic waves, causing cataclysms and panic. It also allowed the newly arrived Archon troops to land and join with the armies of Earth's lower lords in the underworld. This situation threatened all spiritual evolution on Earth, which was in danger of again falling under the domination of the Archons.

The Star Elders' Council had no choice. After several formal warnings, met by refusals from the invaders to negotiate, a decision was made to destroy the Archon mothership/moon by crashing it over the main underworld bases. To blow it up would have caused even more damage, but even so, a major cataclysm was the result, in the eastern part of what is now the North Atlantic Ocean, which brought about the fifth mass extinction on Earth, with the near disappearance of the Saurians, along with major shifts of the land masses, volcanic eruptions and earthquakes.

This cataclysm stopped the Archon invasion of Earth, but a large population remained on Apollyon/Maldek, where new forces could still be brought in through the star gates. In order to protect the solar system from another future invasion, the Star Elders had to eliminate these star gates, as well as put an end to the growing army of fallen souls being enlisted.

Sasquatch Elder Kamooh: "They took a grave and rare decision in the annals of the universe: to destroy a planet with life in order to save a solar system's spiritual consciousness." And so, using incredibly advanced technology, the Star Elders' Council sacrificed the planet Apollyon/Maldek, which exploded and became what is now the asteroid belt between Mars and Jupiter."

The story of "Nibiru" is a past event that is still present in the collective consciousness of Earth-plane humanity. The story of Earth humans being used as slaves to mine for gold occurred later, during the time of Atlantis, when the Repterran "lower lords" were back in power here on Earth, as they have remained to this day. There was a great emphasis on gold in the Atlantean culture, and they travelled to what is now Africa to acquire it, hence the evidence of ancient mines in that region. It was the Repterran/Anunnaki group from Atlantis that was responsible for exploiting slave labor to work these mines.

There is, however, also a deeper meaning here that has caused confusion and misinterpretation of texts referring to off-planet beings coming to Earth to "mine for gold." The deeper meaning refers to the more spiritual-energy transmutation work that the original assistant creator beings were carrying out. They could be looked upon as "cosmic alchemists," as their specific work involves the transmutation of the base "metal" of lower animal/human consciousness to higher levels of divine/cosmic consciousness. This is the evolutionary process that they're continuing to help Earth with right now, as they have with all other human-type cultures throughout the universe. The Indian text, the *Mahabharata*, mentions the Ant People carrying out this work in very ancient times.

NOT ALL REPTILIANS ARE NEGATIVE

It's also important to know that not all Repterran/Anunnaki people down here are Controllers, and even among those in positions of control and authority on the planet, *not all are negative*. Their heritage runs through all the royal bloodlines on Earth. They are very aware of being "different" and tend to feel a sense of superiority and entitlement, but because they've been resident on Planet Earth for so long, some of them are genuinely caring towards the welfare of both humanity and the planet. This off-planet ancestral heritage is the carefully guarded secret held by the highest members of certain secret societies and organizations here on Earth, especially those with roots in ancient Egyptian mystery schools.

Descendants of the Repterran Anunnaki feel a sense of superiority over other newer Earth humans because of this, but really, all Earth humans have an off-planet heritage. We are all descended from the stars. The Repterrans have just been here a bit longer!

Again, let's be clear: God/Source is not a person, but "a high-frequency Energy." It was the resident Repterrans, along with other off-planet people that were the "God" of the Old Testament. They were more than likely responsible for some or all of the following: introducing the Ten Commandments, for leading Moses and his people out of Egypt, the parting of the Red Sea, the mysterious burning bush, and the "manna from heaven" which kept the people alive during their travels in the desert. Unfortunately they were also the ones to use "divide-and-conquer" techniques to keep the various tribes of humans weak and disempowered. The story of the Tower of Babel is an example of this. In other words, they were only human, so the "God" of the Old Testament is prone to the very human weaknesses of jealousy, anger and the need for praise and flattery. As those who have studied the Bible know, some pretty awful stuff went on down here back then, encouraged no doubt by the self-serving Repterran group.

A number of Earth humans carrying a higher percentage of Repterran, Anunnaki and Draconian bloodlines are presently in residence down here as ordinary citizens, not even aware of who and what they are. Being of an older culture, these people tend to be aware of deeper aspects of reality beyond the physical, and are often drawn to astral/esoteric-type studies such as clairvoyance, psychic development, *etc.* The truth is, deep inside, they have a yearning to get back to where they once were as part of an older and more advanced off-world society.

Problems often begin however when such people set themselves up as psychics, healers, *etc.*, because many of them still tend to be rather manipulative and egotistical, and can draw followers in with their very strong power, which they use to impress more disempowered humans. They're very much into playing around with astral energy, which can certainly assist them in seeming to manifest "miracles" at times. Younger humans are easily impressed by such demonstrations of "power." It's important to understand that such "powers" do not necessarily denote a high level of spirituality. The infamous Aleister Crowley, AKA "the Beast," is a good example.

Charismatic Christianity is full of this, which I've had confirmed by two independent sources, one of whom was a Reverend who had been involved in the Charismatic Movement and got out of it because he had awakened to what was going on. The "speaking in tongues" practiced by the Pentecostals is an example of astral-plane Repterran or lower entity interference, as are many magical-type practices. "Speaking in tongues" was originally an ability given to Christ's Apostles to enable them to preach in places where they couldn't speak the language. The "tongues" were legitimate, recognized languages — certainly not the unintelligible gibberish that is now often labeled as "speaking in tongues." This is an example of truth that has been misinterpreted.

During the latter part of the 20th century (1960s) when people began seriously questioning traditional religion, the Controllers could see that they were losing ground in this area, so they then took over much of the blossoming New Age belief system, which was the start of the now widespread drug culture. They influence much of the channeled information given out through psychics, because they're very adept at remote mind control, and in this way have been able to link into the minds of psychics with often fearful messages and predictions to be passed on to humanity. They are also behind some of the Ascended Master and Space Brother cults.

A trick commonly used is to send out false messages to psychics using advanced remote mind-control techniques. They fool some people by appearing as beautiful, angelic-looking beings or passing themselves off as Archangel Michael, benevolent Space Brothers, Pleiadians, Ascended Masters, *etc.* This is *not* to say that all such perceptions come from the Controllers, but many are. Such messages are often fear-based — warnings of end-of-world scenarios and other major disasters that never occur. All the misinterpretation on the Mayan Calendar date of 21st of December 2012 was a classic example, as was the Millennium Bug of the year 2000 that never happened. The key to remember is whether the message is about fear or love. Always feel such a message in your heart and discard any information that feels unloving, hurtful or fearful, no matter who it comes from. Have you ever experienced channelings from some prominent person who starts out with a beautiful message, then once they have a strong following, it becomes a message of fear? Always consider the message — is it love or fear?

Remember the biblical warning of the anti-Christ appearing on Earth as an "Angel of Light." The teacher under whom I trained for eight years in meditation and channeling was very aware of this problem within Spiritualism and other New Age systems. It was for this reason that she didn't encourage the channeling of "angels" or "Ascended Masters" in our group. As she explained, it's just too easy for ego to creep in. As she used to tell us — if a medium has developed long and hard enough to have raised their vibrational frequency to a level through which such highly evolved beings can genuinely communicate, then neither channel nor spirit feels the need or desire to claim such titles. Genuine spiritual masters and angels are very unlikely to proclaim themselves as such.

Many Earth humans are taken in very easily by outer appearance. "Beautiful" equates with "good," and not so beautiful (in Earth-human eyes) equates with negative or even "evil." Therefore, any negative entity out to cause trouble on Planet Earth would of course show themselves as a beautiful angelic-looking being in order to win people over. The Repterran Controllers, particularly those based on the astral plane, are shape shifters, and so can take on any appearance that they wish to.

It's *really important to remember* that authentic channeling is always uplifting and empowering. It is always positive, because genuine higher beings know that we create our reality through our own thoughts, so they would *never* seed fear or negativity in a human mind. They also know that the future is made up of possibilities, never certainties, so there's always room for something more positive to happen. Controller-based predictions are the opposite. They're specifically aimed at causing maximum fear and drama because the Repterran Controllers feed off human fear and emotion, and know that some Earth humans are addicted to drama, and therefore deliberately seek out and are drawn to anything of this nature.

Currently, a massive amount of deception is going on here involving the negative Blond group and their Repterran allies. It's scary just how far their influence spreads on Earth, particularly through religions, belief systems and cults, both traditional and New Age. This misrepresentation comes most easily through people who cannot see past the "tall, blond, blue-eyed" image in which Earth humans have created the "angelic beings." This issue was brought to my attention by a book on past and present ET contact on Earth. It starts off very well with good information with which I can totally resonate, but about halfway through, where the original authors step back and the *co-authors* take over, problems arise. These writers fully believe they're having on-going contact with the Pleiadians, which is possible, but then negative disinformation about the Greys is brought in, with dire warnings about them from the "Pleiadians."

As a former Grey working down here with full conscious awareness of our mission, I have several very dear Pleiadian friends. We work together in total Oneness, with deep love and respect on both sides. We couldn't do this work without each other, and so this "Pleiadian" anti-Grey propaganda is particularly hurtful and upsetting and unbelievable for us Earth ETs. I know there is much deception within the UFO/ ET field regarding such disinformation, and also through Charismatic and Fundamentalist religion. With this in mind, I ran my thoughts and concerns past another good friend who is a very dedicated Christian and faithful follower of Jesus. This friend was also formerly involved with the Charismatic Movement, knows very well how they operate, and has now removed herself from them. She's also been called in on occasions as a "cult buster," to help out others trapped in such situations.

Strangely enough (talk about help from Upstairs!), she too, before I spoke to her about it, had "coincidentally" and recently bought the same book. So when I brought it up with her, she knew exactly what I was talking about and had also picked up on the underlying deceptions within it. The authors may be quite unaware of the depth of deceit in their work, for such is the depth of cunning being perpetrated on this planet. They honestly see themselves as good people and experienced psychics/experiencers who have had past lives as Pleiadians, and they trustingly give out in good faith what they've been told by those whom they believe to be their genuine ET contacts.

The first hint of a problem I saw was the claim by one of the co-authors that his senses had been opened up through psychotropic drugs as are used by shamans, which took him on some amazing off-planet and interdimensional mind travels. He then went on to confess that certain predictions given to him regarding events that were to occur in 2012 - 13, have since been proven to be wrong, but still they persist with their assertions about the Greys being evil. It doesn't seem to occur to them that the "Pleiadian" they put so much faith in seems very egotistical, using the title "Lord" in front of his name. My ex-Charismatic friend recognized this "deception game" straight away, having come up against it on numerous occasions, so I'll let her tell it in her own words:

> "I have often wondered about how much lying and deception there is regarding these (angelic and Pleiadian) beings that folks are channeling and encountering. I have no doubt that some of them are real, and that some truth is mixed in with the messages, but often I find ego, attitude or just plain silliness showing up with it, which for me 'red flags' it as being 'off.' After all, the Bible does say that the anti-Christ will appear as a 'Being of Light', so with such encounters you do need to be very discerning!
>
> "There's so much deception with the whole ET scene, and so much monkey business causing major confusion. Many humans down here, for one reason or another, are purposely or unawares bearing false witness to the whole story. It stands to reason that they're getting a great deal of help in spreading this deception from some of the supposed ETs. I have no doubt that this has gone on quite a bit and that some 'contactees' are being led on, who then get a following of their own as their false messages are spread far and wide."

She then went on to describe a man she had known who was, to put it in her words, a "mega deceiver." He claimed to be a Marine secret ops person, but this was just part of the whole deception game he played. He had definite astral abilities to travel out-of-body and to get inside peoples' heads and be heard, just as the Controllers do with psychics. He'd captured a young, vulnerable divorced woman in his "net," and my friend was involved in a cult-busting operation to get her free of him. The young woman described amazing trips through space with this man, to visit various planets, and much more went on as well. He was involved with prostitution/white slavery/pornography, and was backed by the leadership of a Charismatic Church. It turned out that this leadership was also deeply involved in criminal activities. On several occasions this man made the comment to my friend: "You can sell anything to anyone.... Just sandwich a lie between two bits of truth!" Or, as my friend added — a bit of truth sandwiched between two lies!

Consider the interference — allegedly by the Russians — going on in elections in Western countries, and you'll see how even today "fake news" can sway massive numbers of people. We can see that it continues to be done by mixing some lies with a little truth, and all of a sudden societies are disrupted.

People seeking answers and truth are not stupid. They have sense, knowledge and understanding. What pulls them in is that bit of truth. It resonates within the human heart and soul, and so they keep listening — a bit here, a bit there, and they get lulled by stuff that they want to hear. These deceivers are always so sure of themselves. They have a flow, they're always good talkers, and they're so confident in their own rightness, so people put their questions aside because they feel a resonance with what this person is saying, so they commit to trusting in them unquestioningly.

Soon the deceiver has a whole following, and they're all in there convincing each other about how wonderful this leader is. Peer pressure is involved, along with the need to be accepted as part of a group, the need for reassurance, a yearning for a safe haven and a cozy belief system to cling to and hide behind. Throw in a few astral plane "miracles" — speaking in tongues, channeling beautiful Space Brothers/Masters/Angels or whatever, and introduce an "enemy," preferably alien, demonic or at least non-human as a common "foe" to unite the group into a "them and us" mind-set, and presto, you have a cult!

One of my friend's old professors once wrote,

"Our soul/spirit is our connection to anything outside of our 3-D awareness. It is also our window to Source. If you can reach a person through their belief system, you have them at their core, and if you do this through deception, lies or negativity, you're very effectively 'painting over their soul window'. This then blocks their ability to see clearly and truly because their filters are clouded, or colored, or have blind spots. The little bit of light that does manage to get through is reduced and distorted, so the person's understanding and perception of that light is reduced and distorted. This is a far more cunning and effective way of enslaving a human being than through chains or prison bars!"

Another Controller ploy is to pass on psychic messages of "mass landings or appearances of star ships by the Galactic Federation or High Council of somewhere-or-other." Everyone gets all hyped up about it, followed by disappointment, disempowerment and ridicule when nothing happens, which is always the case. I notice too how such predictions often have an "escape clause" by giving a couple of possible dates! By the time the second date arrives, everyone has conveniently forgotten about it, so nobody even notices when yet again the prediction doesn't come to pass.

The Inner Earth-based Controllers have covert influence over many Outer Earth authorities and organizations. They appear no different from modern Earth humans, and can take their place in society as business, military, government and church leaders, law makers, *etc*. They're often to be found in high positions in lodges, secret societies and mystery schools. The Inner Earth spans both physical and astral planes, because Planet Earth herself is multidimensional.

They possess very advanced technology, including many of the "UFOs" that are seen down here, often around underground bases. In some cases, human pilots have been subjected to aggressive attacks and abductions by these ones, which appear to be "ET" ships but in reality belong to the Controller/Repterrans based right here on Earth.

According to Sasquatch Elder Kamooh, "They have infiltrated your species and societies, running most of your world, as agents of the Archons."

Because the astral-plane-based Repterrans are shape shifters, this can take the form of "negative Greys" in order to cause people to fear the genuine Grey Guardians who are only here to help. For this reason they also carry out cleverly engineered "alien abductions," usually on people who are already having genuine ET contact, to cause maximum fear and confusion. I would go so far as to say that today much seeming ET contact is actually engineered by Controllers/Repterrans to cause distrust and fear of the real ETs, which happens much more than people generally realize.

A classic example was the case of Javier Perez de Cuellar, who, when he was Secretary General of the United Nations, was subjected to a faked "abduction by aliens" which was (to quote Dr. Steven Greer from information passed onto him by Hans-Adam, Crown Prince of Lichtenstein) designed to block an initiative by leading world statesmen meeting at the UN to disclose the truth about extraterrestrial life and technology to the world. Programmed Life Forms (PLFs) that have been designed to resemble real ETs are used to implant abductees with tracking chips to enable ongoing abductions to be carried out on selected people. It was one of these implants that I removed from a human male during work on our ship one night.

The genuine ETs do also use implants, for ease of communication with those of us down here who are on our Earth mission. The implants are also used to monitor our state of health and are generally placed into our chakra system to help keep it well balanced. They assist us in moving through dimensions to board the ships at night as our Earth human body sleeps. The difference between our implants and those of the Controllers is that ours operate at a higher frequency, beyond 3-D physicality. The implants being removed through surgery are Controller ones. I can recall having to cut a wire to deactivate the one I removed from the man up on the ship.

The Controllers carry out hybrid-breeding programs in connection with the military, creating negative life forms such as the Men in Black, and also ones that resemble small Zetas. They're the ones behind animal mutilations. Because they can manipulate human minds easily through remote control, a lot of trickery is going on down here. However, their influence only reaches the astral plane, so, *if people can evolve to higher levels of mind/spirit with the help of the genuine ET Guardians and their "positive" spiritual connections on Earth, then the influence of the Repterran Controllers will be overcome.*

The Controllers' disinformation and negative propaganda is specifically aimed at causing people to fear the legitimate Guardians of this planet, who are here to assist. In the bigger picture, having to work through this deception process does provide a necessary evolutionary test and choice that humans must make at this time — that is, making a free-will choice between love and fear. People must choose, and they can always ask — Is this love or fear? Always choose love.

A typical example of Controller trickery was a message said to have been given to President Eisenhower by a Nordic-type "ET" group that warned of negative gray aliens who were planning to invade Earth. Whether this message really did come from "off-planet blond ETs," or whether this is simply disinformation that has been given out, I cannot say, but it's very typical of Controller disinformation.

The so-called "Blonds" or "Nordics" are an Inner Earth group who inspired Hitler's so-called Aryan Ideal, and drove his obsession to eradicate certain racial types on Planet Earth and replace them with this race of super beings. They were also the reason behind his attempts to find an entrance to their base in the Inner Earth. The Book of Enoch's description of the fallen angels as tall, blond and handsome-looking beings sounds suspiciously like some of the so-called Space Brothers and Ascended Masters of New Age channeling.

The following quote comes from Paul Hellyer, in *Light At the End of the Tunnel,* and is by Sen. Daniel K. Inouye, of Hawaii:

"There exists a shadowy government with its own Air Force, its own Navy, its own fund-raising mechanism, and the ability to pursue its own ideas of the national interest, and free from all checks and balances, and free from the law itself." (p. 46)

It is this group, also known as the Cabal that I refer terran Controllers. Hellyer then continues,

"This highly secret group is, in fact, a 'shadow government of the United States that has usurped powers of the Congress and of the President, and established itself as the ultimate arbiter of 'American security interests.' (p. 203)

"It has its origin in the early post World War II years when America and its armed forces were confronted with the fact of visitors from other planets.

"According to the late Michael Wolf (who is) probably the most extensively involved member of the shadow community to come forward so far:

'There is a group of xenophobic and paranoid generals, charged with the protection of American skies, who fear and hate the ETs and are waging war against them. Called the Cabal, they use Star Wars weaponry including a neutral particle beam to shoot down ET craft and imprison survivors while attempting to extract information by force. The very technology the ETs gave us is now being used against them. Despised by many within the satellite government, this Cabal also uses aggressive methods against those who try to end the UFO cover-up, a concern being that this aggression will intensify as the big announcement draws even closer.'" (pp. 204-6)

Paul Hellyer then goes on to explain:

"This is precisely the reason that I decided to go public in 2005; since then the military universe is unfolding precisely as I had feared. The U.S. generals deeply resented the ETs' interference with their nuclear installations and were willing to spend untold billions of U.S. taxpayers' money in a frantic effort to protect the means by which they could blow up the Earth and make it uninhabitable.

"Although the generals have access to more ET information than anyone else, they absolutely refuse to listen to what the

ETs say, *i.e.*, 'All worlds in the galaxy are interconnected. One Hiroshima atomic bomb can affect every different culture.'"

On p. 207, Hellyer quotes Dr. Carol Rosin, who worked closely with Werner von Braun:

> "Werner von Braun warned us that the military-industrial complex would find one excuse after another to justify perpetual conflict, vast military expenditures — first the communists, then the terrorists, and then the extraterrestrials. The beneficiaries are the new totalitarians who have been running the United States and who wanted to extend their power and influence on a global scale.
>
> "At the end of World War II the United States brought a sizable group of Nazi scientists to America in order to prevent the Soviets, who had recruited a number of their own, from gaining military ascendancy. It was called Operation Paperclip. I have often wondered if some of their mentality hadn't rubbed off on important elements of the U.S. scientific and military communities with whom they worked. Dr. Wolf reports having worked on genetic experiments in one of the great secret underground labyrinths that are virtually identical to some of those that were performed by Hitler's Nazis in the 1930s."
>
> "It is absolutely essential that the people of the United States — Republicans, Democrats and Independents — wrest control of their country from the iron grip of the shadow government, the Cabal, and the "Three Sisters" (Council on Foreign Relations, the Bilderbergers and the Trilateral Commission), as it is inevitable that there are some connections between them — and re-establish some semblance of government of, by and for the people as envisaged by the fathers of their country. (p. 241)
>
> "The situation is now so bad that when Sarah McClendon, a White House reporter, asked President Clinton why he didn't demand disclosure of the UFO phenomenon, he

replied: 'Sarah, there is a government within the government, and I don't control it.'

"What about civilian control of the military over which there have been so many battles? In the U.S. there has been a quiet coup as ultimate power has been seized by a small Cabal accountable to no one but themselves."

Michael Wolf (Kruvant) had an M.D. in neurology, a Ph.D. in theoretical physics, a Sc.D. in computer science, a J.D. in law, an M.S. in electromagnetic influences on organisms, and a B.S. in biogenetics. He was a member of the NSC's unacknowledged subcommittee, MJ12's panel of scientists. In an interview with Chris Stoner, Wolf spoke of meeting with ETs every day in his work, and sharing living quarters with them while doing research at extremely classified underground government research laboratories. He stated,

"Zetas work in underground facilities, as requested by the U.S. Government. The ETs are not breaking the U.S. Government-Zeta treaties, but the Government has broken treaties by mistreating ETs and trying to fire on UFOs. There are some extraterrestrials being held captive."

He once commented that he had never met a Grey whom he did not like. He also told how government scientists had discovered that the ETs cannot dematerialize and escape, if there is an extremely powerful electromagnetic field surrounding them. By way of corroboration, Chris Stoner spoke of a government contractor who described three-foot thick walls with many wires embedded and running through them at Haystack Air Force Laboratory.

Remember to keep in mind that Planet Earth and humanity are in the process of shifting from Fourth to Fifth World consciousness. This Fourth World Earth in which these Repterran Controllers are so comfortably entrenched is coming to an end as human consciousness expands and people begin to question the Establishment — Church, Government and Big Business. The Controllers are desperate to stop this happening because when (not if) it does, they'll lose their positions of control, wealth, manipulation and exploitation of the masses. This major consciousness shift is being assisted by benevolent off-planet beings including the Grey Guardians, who are being assisted by other ETs, whose specific job is to aid evolution in the human kingdom. The Controllers are desperate to stop this happening.

Sasquatch Elder Kamooh: "(But) you must be aware that there are also many lies and deceptions put out by the lower lords who try to be seen as the Star Elders and usurp the title of Earth Watchers (or Hakamin) while they have fallen from their duty (to become Nephilim), due to pride, arrogance, greed for power and selfishness. They use this tactic to be seen as saviors and gods, and carry on their agenda for global control, through a centralized tyranny of their own, that they try to masquerade as the Council of the Star Elders.

"They have encapsulated our home-planet inside a control grid, keeping away the influence of the Star Elders protecting us, through constant warfare, the very backbone of their uncivilization. They have involved the underworld where they conduct in secret their most detrimental and disruptive experiments for the evolution and balance of our home planet, including magnetic and genetic manipulations."

Because the number of genuine off-planet visitors has been stepped up since the end of World War II, the Controllers have used various tactics to diminish their influence on Earth. The first was to take a stance of "official denial," culminating in the infamous Project Blue Book launched by U.S. authorities, which was set up as a way of officially debunking all reports of potential UFO sightings and ET contacts. It was carried through to all media reports dealing with the subject as well, to the point where very few people were willing to come forward because of the risk they'd be labeled as "nut cases" and "gullible fools" for not being able to tell the difference between a UFO and "swamp gas" or some other equally ridiculous claim put out by those sent to investigate.

When too many reliable witnesses such as airline pilots, police officers and even astronauts began speaking out about the reality of off-planet contact, suddenly a number of "military whistle-blowers" came out of the woodwork with stories of negative gray and reptilian aliens who are here to abduct innocent humans and/or take over the planet. This, of course, has caused major fear, and has provided the perfect excuse for the military to build ever more powerful Star Wars-type weaponry, just in case! And so the official denial tactic has been replaced by a campaign of fear-mongering disinformation.

The Repterran Controllers manipulate and enslave Earth humans both emotionally and spiritually, but, while people here allow themselves to be enslaved through materialism, politics, fear-oriented belief systems, wars, advertising, the media, consumerism, fashion, the entertainment industry, *etc.*, there's not a lot that the rightful caretakers can do, as this is a planet of free will, and many Earth humans buy into Controller trickery, which is why many off-planet people are now "coming in the back door," volunteering to incarnate here as Earth humans to try to change things quietly from the inside.

Sasquatch Elder Kamooh: "The lower lords having adopted a strategy of long infiltration with the consent of the victims makes it much more difficult to contain than an open invasion."

The Grey Guardians, along with other benevolent interplanetary and interdimensional people, are here to help free Earth from the Controllers.

Some genuine ET contact naturally involves fear, because Earth humans have to break through many layers and barriers of fear and superstition to enable their minds to open up to a more expanded state of conscious awareness, which contact with the real Greys can engender. Whenever humans encounter the unknown — anything outside their accepted paradigm — their natural reaction is fear. A primary purpose of this book is to educate people about the emerging new paradigm so as to reduce their fear when they suddenly find themselves in it.

The Controllers here on Earth have technology that allows them to actually speak verbally inside peoples' heads, whereas the Greys communicate with us in a different way that's more subtle. It's hard to describe in 3-D terminology. It's not a physical voice, but rather more an "energy signature" that we clearly recognize. I always worry a bit when somebody tells me that they're hearing actual voices, because that's not how the ETs generally do it. If the voice you hear is verbal, then really bring your inner sensing and discrimination into play. If they start ordering you around or playing upon your ego, tell them firmly to *go away*! It's your head-space and nobody has a right to enter without your permission.

It is time to open the doors to the truth of the history of ET involvement on Earth, raise your collective planetary consciousness, and welcome in this new era in which humanity rises of its own accord and takes the reins of its spiritual evolution.

11

Invasion of Free Will

Many people who are undergoing a negative type of contact angrily insist that the Greys are infringing upon their free will. They are adamant that they have never given conscious permission for this process to take place.

If the contact is genuine — and this is a big "if" as I've explained already — then permission has been given. Going back to our Human Ladder, remember that Planet Earth and Earth-human consciousness is only on Level One of the Ladder, and can therefore access approximately 10% of its full potential. The rest of human consciousness is exactly like the greater part of an iceberg that is hidden beneath the surface of the ocean, with only a small tip visible above the surface. All major life decisions and choices, most especially those affecting soul issues such as evolution, are made between lives — they are pre-birth choices and decisions. At this level, a human's higher self knows perfectly well that it's time to start evolving to a higher frequency of universal consciousness, so they can often voluntarily approach the Grey Guardians — the "triers and testers of souls," and request that help be given in the coming life. This is a free-will choice made by the soul that is about to be reborn. You might be asking, "Why the Greys?" Because just the way they look will challenge us much more than a human-looking being would.

Contemplate this for a moment. Can you remember anything from between lives? This request made by the higher self is no different — most people simply don't remember once they've reincarnated back into the density of the three-dimensional physical plane. And even hypnosis generally cannot access such information, as it links in no deeper than the sub-conscious mind, which is the seat of all the fear blockages of past and present, so it is inclined to dredge up a rather distorted account. To be able to access such information, one would need to link into the super-conscious level of mind.

You might be wondering, "If we are at a higher level between lives, why do we need to come back down here to evolve? Why not just stay at the higher level? I understand that beings want to try on physicality, but once experienced, why not remain higher? And what about past-life regressions? Aren't you picking up info from past lives, and can you access between-life states as well?"

Most Earth people don't go to very high levels after they die. They only go as far as the astral plane, and they must keep coming back over many lives because we have to have multiple physical-life experiences to have all our "rough edges" knocked off, so to speak. Evolution (ascension) is not a quick or easy process. It spans many lives, as we work our way slowly up through all the vibrational levels of the Human Ladder. Even when we're ready to move on, getting free of the astral plane is not easy because of certain "illusion screens" that have been set up there through all the various belief systems on Earth. This is why so-called "conscious dying" (page 144) is such an important part of Buddhism. It's to enable the spirit to bypass the astral plane "traps" that keep pulling the spirit back down onto the Wheel of Karma long after it should have moved on.

As for past lives and between lives, again, the astral plane "illusion screens," as well as the fact that a Level-One human mind can only access 10% of potential conscious awareness, effectively stops people from being able to access very much past-life or between-life knowledge. Many Earth humans don't even acknowledge reincarnation, let alone consciously access past-life information.

Some spiritual traditions teach us that we all have higher aspects of our being floating around somewhere "out there" or maybe "up there." Even the Human Ladder illustrates this. So why can't we tune into these higher aspects of self? Some of us can, with sufficient spiritual training and through meditational techniques aimed at self-realization, but it takes a huge amount of dedicated practice to be able to do this, and most psychic sensing doesn't extend beyond the astral plane. The bottom line of spiritual practice is to enable the aspirant to link into higher self-realization, which, in turn, is the path back to Oneness, but astral-plane distractions and illusions can block this. Vickie Mackenzie, in her book *A Cave in the Snow*, cites Buddhist nun Tenzin Palmo:

"The whole point (of spiritual practice) is not to get visions but to get realizations. A realization is the white transparent light at the center of the prism; not the rainbow colors around it."

The problem is that if you get too caught up in "seeing" at the astral level, the "rainbow colors" often hide the center light from your vision.

I am aware that some people reading this are so entrenched in fear and disinformation that no matter what I or anyone else says on this matter, we will not be believed. I'm not denying the fear factor involved in ET contact, including the Greys and Zetas. I, too, was terrified as a child, until I finally woke up and realized that they were not here to hurt me. I began to understand what was really going on, and all I can say is, there are many others out there who have also awakened.

No genuine Grey Guardian or any other ET person ever overrides human free will. The Universal Law of Free Will is one of the 11:11 Universal and Spiritual Laws of the Cosmos by which all must abide. (page 170) These laws, however, are not always practiced on Earth, and until they are, Planet Earth will not be permitted to become a universal member of the Star Nations.

12

Humanity's Call for Help

Over the past 70 years or so since the first atomic bomb was detonated on Earth, the off-planet Guardians have stepped up their program of contact. This negative event on Earth ran totally contrary to Universal Law but there were still positive aspects, as there is with even the most negative-seeming event. Firstly, it effectively brought WW II to an end, and secondly, it caused reverberations of horror around the entire planet. This sense of horror in the heart of humanity was the catalyst that sparked off the necessary recognition in the collective human psyche that the time had come for the next step in the journey of human evolution. This was the free-will call for assistance from humanity that the off-planet group had been waiting for.

Evolution is always driven by the need for change — the need to shift existence from the ocean to dry land, the need to develop lungs in order to be able to breathe out of water and the need to adapt to walking on two legs rather than on four. The step that came after that for the human kingdom was the need to develop reasoning power in order to exercise free will, and the next step upwards, which is happening right now, involves coming to fully conscious self-realization of the God Source Energy within each one of us. This understanding enables a human being to evolve from animal human to cosmic human by choosing love over fear. Once this next evolutionary step is achieved on Planet Earth, warfare will cease to exist. At that point in time the human species of Earth will earn their right to cosmic citizenship, and it is the task and privilege of the creator beings to assist this process, as they have with countless other planetary cultures. Although it may not seem so right now, people are waking up. It will happen, but it all takes time.

Evolution is never an easy or quick process, even under the best conditions, and when it involves a human species on a planet like Earth, where conscious awareness is on the very first level of the Human Ladder, it requires the breaking through of many, many limitations and blockages of mind. These limitations and blockages have built up in the Earth-human psyche over multiple generations of fear-based belief systems, superstitions, misinterpretations of wisdom teachings and a massive amount of disinformation that has been spread consciously by a few and then unconsciously by many.

It amuses the off-planet people greatly when Earth humans are led to believe that their Ascension process will happen in one split-second, blinding flash of light, as some believed would occur on December 21, 2012. I repeat — nothing is supernatural — everything is governed by Natural and Cosmic Law, including evolution, which is a very slow process on a Level-One planet. Earth humans think it's not happening because it's taking so long, but time as it's measured on Earth is not real. It is simply an invention of 3-D limited minds.

People often ask why the ETs don't land openly "on the lawn of the White House." The answer from their viewpoint, is that they are not here to prove their physical 3-D reality. They've already done that, and the evidence has been hidden away through no fault of theirs. However, their main purpose for coming is to help you to open your conscious awareness to higher realities. To repeat, evolution is always driven by a need for change and adaptation. This is the catalyst that drives evolution in any species. The greatest challenge for Earth-plane humanity at this time is to expand your conscious awareness, to break through the "glass ceiling" of perceived 3-D reality in order to enable higher perception beyond the limits of that reality, which is why they seem covert and secretive to many of you. It is to encourage you to open your inner perception and higher awareness, which is your only way "Home," to borrow from Christian terminology.

Evolution is happening, but at this point in Earth-plane time, not everybody is quite ready to take the step yet, so a lot of polarity is occurring, with the Controllers putting in an extra effort to pull people back down into fear. There are many Earth humans who are still fence-sitting, not yet able to make a clear choice between love and fear, which is not easy. You can't have it both ways, and everyone must choose in the end. In this regard, I am reminded of an amusing observation made by a Grey Elder that perhaps fence-sitting is why the Earth-human posterior is shaped as it is! And who says they don't have a sense of humor?

The rise of Atlantis has been long-predicted by psychics for the end of the 20th century, and that is precisely what has happened, but in the case of Atlantis, it's not the physical land mass that has been reborn, but rather the spirit of the land and its people. Unfortunately you can see similar attitudes in what is happening on Earth today — especially in the Western World — is a repeat of Atlantis. Reincarnation is a spiritual/soul process— it is not the physical body that rises again from the dead but rather the spiritual aspect. Let us hope that this time love will prevail over fear. It is up to us in our power as humans of Earth.

Development of Earth-Human Spirituality

*"Darkness cannot drive out darkness;
only light can do that.
Hate cannot drive out hate;
only love can do that."*

— *Martin Luther King, Jr.*

13

Distorted Spiritual History on Earth

Volumes have been written on the development of the human race on Earth, often with glimmers of truth buried beneath many layers of confusion, misinterpretation and disinformation.

Walter Benjamin once wrote, "History is written by the victor." In other words, modern-day researchers are only able to access one side of the story, and that side is most likely biased, embellished, and therefore, not necessarily accurate.

Added to this is the problem that once we start talking ancient history and quoting scriptures and texts dealing with it, like the Bible, things get even more complex and confused. The Elders taught us to look at everything on three levels and this is a vital point that is often missed. The fact is that most of these ancient writings are actually composed of three distinct levels. A good illustration is the three levels of Jewish scripture, the Torah, the Talmud and the Kabbala. The first outer level, the Torah, takes the form of a parable, which could be likened to a story suitable to be read to a child. This outer level most certainly contains valuable life teachings and morals, but these are greatly simplified and set in a way that young souls can link into and relate to, just like a children's story. This level is aimed at the emotional side of the human psyche.

The second level, known as the Talmud, is hidden a little deeper beneath the surface and it can be comprehended only by studying the text with a more educated mind. This level is aimed at the intellectual side of the psyche. To get the most out of study at this level, the material must be approached with a mind that is pure and clear of preconceived ideas and prejudices, which tend to cloud true understanding of what the original author was trying to convey. Unfortunately, here on Earth, many translators of ancient texts have not done this, but instead have fallen into the trap of introducing elements of their own mind-set into their interpretation, thereby obscuring the original meaning.

For example, Mark D. Siljander, author of *A Deadly Misunderstanding: A Congressman's Quest to Bridge the Muslim – Christian Divide,* is a linguist who read the Qur'an, or Koran, in Arabic, the language in which it was written. He also read the New Testament in Aramaic, and learned that nearly all the misunderstandings between Muslims and Christians are due to the Greek and then English texts that distort the original meanings. Siljander spends most of his time in constructive debate with Muslim and Christian clerics and scholars.

The Aramaic and Hebrew languages, like many ancient tongues, are multidimensional, with several layers of meaning within, whereas the modern English language is limited in this sense, which is why Siljander found the English interpretations of the Bible to be inaccurate and limited in their interpretation from the original Aramaic. The English interpretation has lost much in depth and correct understanding of what was originally being conveyed. As an example, please read the original Aramaic version of the Lord's Prayer (page 123,) which clearly illustrates the difficulty of trying to translate higher and finer multidimensional reality into the limited confines of 3-D reality.

The third and deepest level of understanding, the Kabbala, is aimed at the soul aspect of the psyche, and it can only be accessed by moving beyond emotion and intellect, into contemplation and meditation. Symbolic language has been used in all of these ancient writings, the inner meaning of which can only be unraveled through many hours of meditation involving a deep understanding of what we refer to as the Source, or what Earth humans call "God," which is expressed through Universal Consciousness/Life-force Energy. The use of intellect alone at this level can cause a huge amount of confusion and misinterpretation, because the concepts involved are generally multidimensional and therefore way beyond the three-dimensional level of consciousness at which most human brains operate. At this level one must embrace simplicity and have a very open and clear mind.

The history of Earth and the development of the Earth-human species are very complex, and numerous writers and scholars have proposed many scenarios. The problem is that such complexity can be manipulated by mixing a little truth with much fiction or surmises to cause confusion in peoples' minds. In today's terms, this process is called "fake news."

A typical example is the concept of humans having been created in the image of God. Many Christians have been led to believe that only Earth humans have been created physically in the image of God. They then take this misconception further by claiming that the Earth-human form must therefore be the most perfect and Godlike form in the universe, and, if other off-planet beings such as the Zetas and many other strange-looking beings do exist, then they must be either angels or demons, because they don't look human in the Earth sense of the word. Or they believe that humans are at the peak of Creation on Earth and all other Earth life forms should be controlled, which has caused our ecosystems to become off balance. This view is a totally incorrect assumption that has grown out of a badly misunderstood and misinterpreted truth.

The origin of this misunderstanding is the fact that the Earth-human physical form does resemble that of the Anunnaki, a reptoid/human species to which Earth humans are closely related. Early Earth humans were purposely led to believe that the Anunnaki were "God," which is just one such misunderstanding and misinterpretation in the Bible, which has persisted over many centuries, causing major problems. Please note — I am not claiming that the Bible is wrong. The Bible contains great truths and wisdom. It's people's interpretations of some parts that cause problems.

Human form throughout the universe always follows the same basic pattern of walking upright on two legs, and having two arms and a head, but there are many variations, depending upon the animal species from which a particular human planetary culture has evolved.

As we discussed earlier, today's Earth humans were developed from a primate species, which is evident if you look at the skeletons of earlier humans. Human life on other planets also shares certain similarities with the animal form from which they evolved — for example the Zetas' soul containers originated from insectoid stock, and they still retain the thin body and in some cases the chitinous exoskeleton and compound-eye structure of some insect species.

There are human types on some planets with catlike features like eyes with vertical pupils, and others who are winged like birds, or who possess the gills of fishes or webbed toes and fingers of amphibians. However primitive many of these other non-primate humans may appear, they are way ahead of Earth humans on the evolutionary ladder.

14

Ancient ET Evidence in Australia

From early childhood I've always felt a strong connection with Australia — a primal, inner soul link with the land itself rather than simply a nationality connection. It's always been as if the land is a living being whom I feel the need to embrace and meld with. This feeling is difficult to put into words, but I know that it is shared with our indigenous people — the many Aboriginal clans who have called this country home for many millennia before Europeans set foot upon Australian shores.

During the eight years I spent in a meditation circle, this inner connection became clearer as past-life memories began surfacing of a time spent in what New Agers refer to as "Lemuria." I've never been able to link comfortably with "Atlantis," but Lemuria has always stirred a similar sense of connection as does Australia. When this past-life link came up, it reminded me of a memory/vision I used to experience on a regular basis when I was about nine or ten years old.

This memory was of a place where we wore long, white robes tied at the waist with various-colored sashes. The color of our sash depended upon the work we did and the group to which we belonged. This is also a reflection of Zeta society, where one's type of work and place of abode are closely linked. For example, all disc pilots live in one communal area, all scientists in another, healers in another, *etc.* I felt that my vision reflected an off-planet arrangement that we brought with us to Planet Earth. Many years later I learned that Australia was part of Lemuria, and perhaps these memories hark back to that time.

One of the most sacred sites for the Aboriginals lies at the very heart of our country, the "Red Center," which is known to most overseas people as Ayers Rock, named for Henry Ayers, a South Australian government official in the 1870s. This site, along with the many thousands of square kilometers/miles surrounding it has now been officially handed back to its rightful caretakers — the Pitjantjatjara people.

Like most Indigenous peoples, the Aboriginals don't look upon themselves as owners of the land. To their way of thinking it's the other way around — the land owns them — they're simply the caretakers, and their priority was to restore "Ayers Rock" to its sacred name of Uluru. Nearby lies another very sacred site, formerly known as the Olgas, in honor of a European Queen, but now restored to its traditional name of Kata Tjuta, which translates into English as "many heads" (it's a small mountain range). Kata Tjuta is also the name given to the huge Central Australian National Park in which these sites are located.

Part of Kata Tjuta has been leased back to the government to allow tourist resorts to be built, but visitors are asked to not climb Uluru, as it is a sacred site. To climb it is considered extremely disrespectful and also quite dangerous. A number of people have died in the attempt, usually from heart attacks, as the temperature there is often up to or even over 40° Celsius (104° F), and strong desert winds blow. Tourists are also asked to not take home rocks or pebbles from Uluru as souvenirs, as this is a taboo that brings misfortune down upon those who do. There's a "sorry book" set up in the local Cultural Center with a pile of these stones around the base of the book stand. These stones have been returned by people affected by the taboo, which can only be lifted by bringing them back to where they belong and writing a formal apology in the book. The book records many tales of the bad fortune that has befallen people who have taken as souvenirs these small, seemingly innocent, pieces of Uluru.

All my life I've wanted to visit this place, and my chance came in September 2013 when Helene and I, along with another like-minded friend, decided to make the journey. This decision, like my trip to the UK in 1995 (see p. 64 of *The Zeta Message*) where I experienced an evolutionary jump, also set off some very interesting events. When I told Sanni Ceto of our plans, her reaction was: "Oh, yes! That is where you were heading when your ship crashed. We have a connection there." This information from her came just when I was writing the section in this book dealing with the Creator Beings coming to Earth, and I knew it was no coincidence. This was confirmation that I really needed to make this trip.

Uluru and its surrounds are steeped in "Dreamtime" legends of the Rainbow Serpent who, along with the off-world Creator Being ancestors, created Planet Earth and all living creatures including humankind. As far as I understand, the Rainbow Serpent is symbolic of the two polarities of life-force energy — yin and yang — negative and positive — which the Creator Beings tap into in order to create life (God/Source Energy in other words). It is also symbolic of DNA. The Aboriginal people honor and fully acknowledge the star people, just as the Native Americans and many others traditionally acknowledge their Star Nations ancestry. The Pitjantjatjara people of Uluru, and, in fact, many Aboriginal people, trace their ancestry back to the Pleiadians.

AUSTRALIAN ART

The term "Dreamtime" or "Dreaming" is not an accurate translation from the Aboriginal language, which contains many complex spiritual concepts, but it basically means a deep spiritual state when contact occurs with the ancestral Creator Beings (ET contact experiences). It is not a past or future event but rather a continual *Now* happening. It is said that after creating the sacred world, the spiritual beings turned into rocks or trees or parts of the landscape, all of which became sacred places. The spirits of the ancestor beings are passed on to their descendants. The Creator

Beings are known as Wandjinas, which are described as having large eyes, like the eye of a storm, but no mouth. The resemblance to Greys is quite obvious.

According to Valerie Barrow's book *Alcheringa*, Alchquarina was the name of the Lyran Commander-in-Chief of a Pleiadian star ship, but over the span of many millennia it's become "Alcheringa," with its off-planet Creator Being and Dreamtime connotations.

These Beings are held in such a place of honor and respect that only certain highly initiated artists are allowed to paint them, and they feature heavily in Aboriginal art. Wandjinas are always drawn with an aura or halo around their heads, as angels are depicted in Western religious art. The presence of these Beings is very strong at Uluru and Kata Tjuta. Many faces are visible as natural features on the rock face; everywhere we went we were aware of them, and their expressions change with the shifting sunlight and shadow. Similar depictions of Greys have also recently come to light in Mayan art.

MAYAN ARTIFACT DEPICTING A GREY STANDING BENEATH A CRAFT

Interestingly, I only learned afterwards about the Aboriginal legends concerning the Creator Beings and how they place part of their own essence into the landscape. Hearing this was confirmation to what we felt at the site. We spent as much time as we could at both sites, Uluru and Kata Tjuta, and the energy was amazing — yang energy at one, yin at the other.

After rising early enough to welcome a magical sunup, we were drawn to make one last visit to the Rock. The day before, at a place known as the Mutitjulu Waterhole, we'd discovered an amazing natural feature high up on the rock face. We could see it clearly from ground level, and it appeared to us as a very distinct face and body of a Grey with its right arm extended (directly above the large central fissure).

As we spotted it, a massive wave of energy flowed over us. This figure and the other faces visible on the Rock seem to support the Aboriginal legend of Alchquarina and the Creator Beings placing part of their essence into the landscape. If, as Sanni said, we have a connection there, this also provides a basis for the legends linking the Creator Beings to Uluru. But what I still wanted to know was, what was my personal connection, and why had I been traveling there when our ship crashed in North America?

We retraced our steps to find the Grey figure, and once again we were dumb struck by the incredible energy and depth of wisdom and kindness in the facial expression. Remember, this is not a painting — it's a natural feature untouched by human hands! An urge suddenly came upon me to walk back a little way on the path we'd followed where I'd noticed an almost invisible cave tucked right in under the Rock and screened by vegetation. Leaving the others to contemplate the Grey image, I made my way off the path and through the undergrowth. It was still early, so no busloads of tourists had arrived, and the place was totally deserted

All the caves around the base of Uluru are sacred sites where Aboriginal ceremonies are carried out. Sometimes the whole area is closed off to tourists while a ceremony is being conducted, so if any tourists had been present I wouldn't have gone into the cave so as not to draw attention to the fact that it was there. With the sacredness of the area in mind, I approached quietly and respectfully. I knew that I was being called into a place that was potentially taboo to the uninitiated.

I sat down on a large stone in the cave and placed my palms against it in the ET way of creating an energy link, at the same time

NATURAL ROCK FORMATION AT ULURU

sending out a telepathic message of love, connection and gratitude for the privilege of having been allowed to enter. The whole experience was quite overwhelming. As I quieted my mind, Maris began sending messages to me inside my head, telling me how he had visited this part of the planet many, many millennia ago and of the deep love he'd felt for the place. Tears came into my eyes as I experienced the depth of connection he'd felt with this ancient land. Then came the time when the Creator Beings were driven from Earth by the Repterrans and he was forced to leave. As a Creator Being, he was able to place part of his essence inside the Rock, so that even though he had to abandon the planet, part of his being would remain here forever.

When we blended soul Zetas returned to Earth in the 1940s, part of our mission was to visit Uluru in order to reconnect with the indigenous people of the area, and also for me to re-establish a link with this higher-soul aspect of my own being — the one I know as my Elder/Teacher, Maris.

Remember, the Human Ladder is also symbolic of our Higher Self, and aspects of that Soul Self are manifesting on all ten levels. I never made it to Uluru back then because of the crash, so I chose instead to be reborn in this country several years later, in 1952. My childhood attachment to Australia is very deep, because it carries the reincarnated spirit of Lemuria. Being born Australian in this life has made my return visit to Uluru easier to achieve.

KARIONG

Further awareness of a deep and abiding past-life link with Australia came to light when I was put in touch via a mutual friend with Valerie Barrow, which adds another powerful Australian Aboriginal link to this story.

Kariong is the site where the ET rescue pod crashed into the ocean after it was ejected from the attacked Pleiadian mothership Rexegina. (See page 44.) In addition, it contains a huge amount of independent validation that has been provided by many people, and third, it presents the Genesis Story with respect and acknowledgement towards the Aboriginal people, whose ancient lore is based upon the off-planet Creator/Ancestor Beings who were responsible for the seeding of Earth and the development of the Earth-human species. Alcheringa tells the story of a group of these off-planet colonists whose star ship crashed on the east coast of Australia many millennia ago.

Astonishingly their story has been recorded in stone on the walls of a cave at Kariong to the mystification of archaeologists.

This area can still be visited today and Helene and I had the opportunity to do that in October, 2016, right after we spoke at a UFO research group in Sydney, New South Wales. We were joined by two star seeds from the United States, whom we met at our presentation, which was focused on speaking out openly as Greys carrying on their mission on Earth. This was quite a challenge and a huge breakthrough for us. Our talk went very well, the audience was friendly and accepting, and we finished with a DNA-enhancing and chakra-balancing meditation that Helene had been given by one of our Grey Elder Teachers.

The day after our presentation we travelled north to Woy Woy, near Gosford, to meet with our guide, Nina, who was to take us to the Pleiadian and Egyptian glyphs at Kariong. Many months before this trip was planned I'd had a dream about visiting this place. In the dream we travelled through two inter-dimensional portals, and entered a narrow gorge where hieroglyphs were carved on two walls. I was shown that Helene had been one of the artists who'd carved the symbols. Remembering this dream, I was anxious to see how I would feel at the site itself.

Nina met us at the train station and we headed out to the Bulgandry Aboriginal Site. The Bulgandry site is dedicated to Baime, the first Creation Spirit of Aboriginal lore. On top of the ridge is a rock platform on which there's an engraving of Baime with one of his arms pointing directly towards the Kariong glyphs. The legend tells that right back at the beginning, after the animals of Earth had been created, there were no people. Baime then decided to make the human species, but these creations quarreled, fought and showed no respect. Baime wanted to destroy them, but his feminine counterpart objected, pointing out that the humans needed guidance. Baime has no gender, but this original Creator Spirit appeared in male form and presented the wisdom of the ages to humanity.

Before entering this sacred area, we paused for a short ceremony to honor the Guardian Creator Beings and to ask permission to enter. Nina, our guide, has undergone many years of training under a highly respected Aboriginal Darkinjung Elder, Aunty Beve Spiers. Aunty and Uncle are terms of respect for Elders used among the Australian Aboriginal people. This Wise Woman has transitioned since passing the sacred lore onto Nina, but she's always present in spirit when Nina carries out this work. We acknowledged her by carrying a white cockatoo feather to honor her clan.

Being on a spiritual quest, we honored everything that came on our path! Two important plants we saw were the Grey Spider flower and Bush Fuchsia. Because Helene is a practitioner of Australian Bush Flower therapy, the special healing properties and the significance of them appearing on our path did not go unnoticed. Grey Spider flower is used to release deep fears and terror, and to assist with faith, calm and courage. Bush Fuchsia represents intuition, clarity of speech and integration of information. Given the experiences we were about to go through, the appropriateness was truly appreciated!

SCRIBBLY GUM TREE

The area is also thick with "scribbly gums." A type of Eucalyptus with smooth, light-colored bark decorated with the artwork from a small burrowing creature. Some of these "scribbles" were amazing, closely resembling human-looking figures and faces as well as animal-like forms.

Faces can also be seen in rocks as we observed at Uluru, underlining the Aboriginal belief of the Ancestor Creator Beings having taken up residence in the landscape, as guardians of the land.

GRANDMOTHER TREE

At one point we met another gate keeper - a small Bearded Dragon lizard who stood boldly in the middle of the path, but who graciously stepped aside and allowed us to pass.

The next gatekeeper was a giant swarm of native bees filling the entire path with a buzzing cloud. As we approached they too parted for us. The fourth wonder we encountered was a magnificent lilac Wisteria vine flourishing in the depths of the Australian Bush, untended and wild, in full flower. Wisteria symbolizes openness, nurturing, creativity and gentleness. In other words, the yin side of Creator Energy.

After what seemed like hours of walking, we arrived at the Grandmother Tree. This is a giant Angophora tree embodying a female Elder/Creator Spirit who guards the entrance into the gorge containing the glyphs. We paused to rest and honor Her before continuing up to the site.

Spiritual quests are never easy. Their whole purpose is to present a challenge designed to strip the mind back to the bare bone in order to reveal the shining diamond of the pure soul essence within. Having been through such an experience myself, and remembering the dream in which I saw Helene as one of the Pleiadian glyph engravers, I knew it was now her turn, for which she'd been preparing all year. It was to be a major Rite of Passage for her.

This quest began deceptively easy, presenting us with a slight uphill ascent over smallish rocks.

As we progressed, there was a subtle change in the energy, and next moment our first test presented itself in the form of a tiny opening that we had to climb up to, then bending double, squeeze ourselves through. A sense of protectiveness and *déjà vu* came over me, urging me to allow the others to go first to make sure they made it through safely. But Helene held back, wanting me to go ahead of her. This sense of "been there, done that" was weird, because I've never personally related to the crash of the Rexegena Pleiadian ship that crashed here nearly a million years ago.

However, recall has since emerged regarding the circumstances behind this *déjà vu* experience. I was one of the Ant People living in the area where the Rexagena crew came ashore. Among them were the ones who are now Helene and Nina. My people gave shelter and protection to the survivors, getting them out of the harsh sunlight by taking them down into our deep underground caverns. All but the most basic of their medical supplies had been lost with the ship, so when it was time for them to begin their genetic work, we provided them with the necessary equipment and a suitable place to work, along with our own expertise in this field. Helene was a Lyran crew member who'd carved some of the glyphs, hence the deep catharsis she underwent while retracing her steps and coming face-to-face with her own handiwork this time. Back in that past life, making the climb up to the entrance of our tunnel was a massive challenge for the injured and exhausted crew members. This time Helene again worried about whether she could make it or not, but just as it was back then, she knew she had to.

Once Nina and I were safely through, we encouraged Helene to crawl hand-over-hand through the tiny opening which was barely two feet in width and height. Once through, the next challenge was a very difficult scramble up, around and through another smallish opening to reach the narrow gorge where the hieroglyphs are situated, Pleiadian on the left, Egyptian on the right.

ENTRANCE

By this time we were struggling with the same feelings that Valerie Barrow describes the marooned ETs suffering — gasping for breath, physical weakness and uncertainty as to whether we'd make it — or not. But again there was no choice, no turning back. We had to make this final ascent.

Again Nina went first and I followed, handing my backpack up to her while I crawled up, searching for foot and hand-holds in the almost sheer rock wall. It was an awkward climb with twists and turns and a final scramble up and over a sharp lip of rock. Then it was Helene's turn. It took her several minutes of getting past the fear barrier and into the right head-space to make this major ascent upwards and over into the gorge. It was the biggest challenge of her life. Nina and I kept talking and encouraging her, and I put Reiki symbols over her to give her strength and courage. It wasn't just a physical challenge but was emotional and spiritual as well, because as she neared the top, the very first thing she saw was an engraving of an ET ship — an engraving I'd seen in my dream, that she had carved herself all those millennia ago. Gasping for breath, shaking with exhaustion and with sobs coming from deep within, she fell forward to place her hands on it. It was obvious that she was re-living all the pain from the past, processing, healing and clearing it from the depths of her soul.

After allowing Helene the time and space needed, Nina then led us up through the gorge towards a rock ledge from which could be seen the area at sea where the rescue pod had

First Vortex

crash landed, but challenges always come in threes. There was still one more rock wall to climb, as high as our heads or more, with no foot-holds and slightly concave so we couldn't crawl up it. Shaken and exhausted as she was, Helene was determined to find a way. She suggested we wedge some rocks in halfway up to provide us with a step. We were soon up and over, looking out to sea to where the ship had come down.

Second Vortex

Next Nina led us to a second vortex, the first one being situated just beyond the opening into the gorge. We later found that this first vortex had magically appeared by itself on Helene's camera between two photos she'd taken in quick succession to record and celebrate her achievement of getting through the entrance.

The second vortex, after the gorge, is marked by two circles side by side in the rock. Helene and I sat down together, one of us in each, to meditate for a few moments. She "just happened" to glance at her watch and found it to be right on 12 noon. This was the exact time that I'd stood in the center of the circle at the Rollright Stones[1] in Oxfordshire, UK, in mid-June of 1995 (the day of the Summer Solstice). This was where my Rite of Passage had taken place. We then became aware of being encircled by a group of Aboriginal and ET people moving slowly around us in an anti-clockwise circle, activating and sealing the energy.

We went back to join Nina, who was very excited by the engraving of an arrow she'd found in the rock underfoot. In all the many times she'd brought visitors to this spot, she'd never noticed this arrow. A message came through that it pointed to the exact spot where the rescue pod had come down.

There was one more thing that she wanted to show us. This was a massive rock with a hollow carved out of the center, known as the Birthing Stone. It's believed that this stone was used by the ET females who, through surrogate motherhood, were attempting to produce a more humanized species to upgrade the "mixture" species that the Repterrans were exploiting for slave labor. You'll recall that the ETs were attempting to carry these hybrid embryos in their own bodies, and the Birthing Stone was where they came to give birth. The attempt was not successful and caused major heartache for the ETs, whose whole reason for coming was to assist the new human species to evolve to a higher energy frequency through an infusion of off-planet genetic material.

---

1. An energy essence extracted from a crop circle at the Rollright Stones can be found at http://notmadebyhands.com/2017.html#1739

Nina has brought many people to this place, and the Birthing Stone is huge, but on the day of our visit she could not find it. She searched high and low, tramped all through the Bush where she knew it was located, but it had vanished into thin air. Apparently it was not meant for us to see it, and I could see that Helene was relieved as well. Some memories are best left to recede into the past. Humanity on Earth is evolving, slowly but surely. It is now time to move on.

Nina has since contacted us to let us know that the following weekend she took another tour group there, and the Birthing Stone was back in place, right where it was supposed to be, exactly where Nina had searched and not found it. It's a massively large rock, impossible to miss. The Greys are certainly clever when it comes to protective screen imagery!

GLYPH OF A PREGNANT STAR WOMAN.

KARIONG AND THE EGYPTIANS

Around 3,000 to 4,000 BC, Egyptians came to Kariong on pilgrimages to pay honor and respect to this original place of learning and wisdom. The gold used in Tutankhamun's tomb carried an Australian signature and was sourced at Ballarat, in central Victoria, Australia. A dozen wooden boomerangs with traditional Aboriginal patterns were also found in his tomb.

The oldest Egyptian hieroglyphs are right here in Australia, and they engraved their own deities and iconography in the gorge alongside the Pleiadian engravings. Here there are also hieroglyphs telling the story of two brothers who arrived by boat from Egypt on a pilgrimage to this ancient site. One of them (Nefer Ti-Ru) met his death in the Australian Bush after being bitten by a snake. They'd come because they knew that this place was the sought-after "Garden of Eden," where animal man was genetically enhanced to become human by off-planet beings who were trying to intervene after the hijacking of the planet by the first-born Repterrans, and where ancient star knowledge had been passed down to the indigenous culture from advanced star beings who had carried out this work.

These Repterrans considered the planet as their possession to be exploited to the fullest, and the newly developed "mixture" species as mere slaves to be used and abused. The original people, the Aboriginals, have proto-Egyptian hieroglyphs in their languages. Many believe that Egyptian hieroglyphs were originally based on Pleiadian hieroglyphs, and the similarities can be seen at Kariong.

Isis Grey

15

The Three Levels of the Adam and Eve Parable

The Adam and Eve parable is another example of a story on different levels. One level of the Genesis story tells of Eve being created out of Adam's rib. At this level, Adam and Eve are representative of the first humans, the original people created or developed on Earth by the Assistant Creators. Adam is symbolic of the first prototype of *Homo sapiens sapiens*. Genetic material was then extracted from this prototype (Eve taken from Adam's rib) to develop more of the same type of being. The prototype was actually female, with XX chromosomes, so in reality "Adam," carrying XY chromosomes, was a mutation taken from "Eve's rib."

The second level describes Adam being created out of "clay," which is symbolic of the higher-dimensional assistant creator beings coming down to Earth to carry out the task of developing a human species native to the planet. They had to take on physical human form, a form of "clay," to enable them to do this. Eve being taken from Adam's rib is symbolic of the Earth-human race being created out of off-planet genetic material. Because of this sharing of essence, there was a strong energy link between the two, with the higher beings looking upon these newly developed human beings as their "children." This part of themselves needed to be nurtured and protected. It is considered a great honor to be given the role of creator and nurturer of a new human species on a planet.

This is why a place of safety and protection, or the "Garden of Eden," was set aside for them. This was the headquarters of the rescue and development program that was being carried out. By placing the new species in this safe environment, the ET Guardians could keep them from mating back with the older hominid species from which they'd been developed, thus preventing any chance of devolvement occurring. An "energy refinement process" was put in place, which was more than a genetic process. The Assistant Creators also carried out work on the energy systems of the new species, placing implants into their chakra systems to assist with gradually and gently raising the vibrational frequency of their energy bodies. This process is currently being carried out again, using implants and other means such as the energy encapsulated within authentic crop circle patterns (see page 198) to assist Earth humans with their "ascension" to higher-energy frequencies.

One way the Repterran Controller version of events has been promoted was through Earth-plane religions. Other ways were through governments. The story of Eve being the one to tempt Adam into committing the "original sin," is in reality a misinformed version of "the Sons of God mating with the Daughters of Men." The "daughters of men" were the female offspring of the new human species. The Repterrans took no responsibility for the act, but rather blamed their temptation and subsequent downfall on the female humans. This happens today, especially in cultures and religions that are still strongly under Repterran control and therefore overly patriarchal.There are still some belief systems today in the 21st century that refer to the female gender as being "sinful" and less spiritual than their male counterparts.

These Fundamentalist belief systems still promote the idea of a weak and sinful "Eve" leading her righteous male partner into the forbidden act of sexual intercourse, thereby being banished by "God" from the Garden of Eden. However, in practically the same breath, God orders humans to "go forth and multiply."

The third and deepest level of the Genesis story is symbolic of the opposite polarities of yang and yin, positive and negative, male and female. The further down the Human Ladder we descend, the more polarized into duality we become, including the genders of male and female. Oneness as it's known and expressed on higher levels of human consciousness simply cannot be expressed in the denser and more solid 3-D matter frequencies of a level-one planet like Earth. In this sense, Adam and Eve are symbolic of the opposite polarities of positive and negative as expressed at this level — not negative as in bad or evil, but rather negative as in electrical polarity.

This extreme and dense Earth-plane duality and polarity counters the argument often raised in regards to negative ETs. Because there are both good and bad people on Earth, then surely with the Greys and other ET cultures the same would have to apply. But it doesn't. In higher planetary cultures, negative and positive work together in Oneness, so there's less separation and more empowerment. Because of Earth's extreme duality and polarity, humans experience disempowerment, which is precisely why Earth is the number-one "school planet" in the universe. Its inhabitants have enormous potential to exercise the free-will act of choosing right or wrong, and therefore maximal opportunities to either evolve or devolve. In other words, Planet Earth really sorts the sheep from the goats. Many other level-one planets don't have this much free will.

Other universal cultures, including the Greys, have evolved to higher levels on the ladder. Being much more homogenized and attuned to Oneness and God's will, they cannot be compared with Earth humans. Because of this enormous polarity and duality, Planet Earth is also known widely as a war planet, to be avoided by most ET people from higher levels of the Human Ladder. They look at those of us who have come here voluntarily to assist as being very brave and heroic! It is a difficult assignment.

16

The Lord's Prayer from the Aramaic

To illustrate how translations can change, I present the beautiful prayer known as The Lord's Prayer, translated by Dale Allen Hoffman directly from the original Aramaic language spoken by Jesus.[1]

This Aramaic Prayer of Yeshua is from The Gospels of Mattai & Luqa and you will see the difference between The Lord's Prayer we've been taught when you read this experiential interpretation by Hoffman.

> Our One, Absolute, Eternal Being, of which we are borne forth from the Realm of the All and the Only
>
> I am empty within the ecstasy of Your Presence and the purity of the vibration of Your Name
>
> Empower my creative expansion through Your emergence from the unseen realms
>
> As I realize our life and will as One
>
> On the manifest Earth as in the un-manifest heavenly realms
>
> Provide the nourishment of authentic insight and realization through me now and in each present moment

---

1. For more translations of ancient texts from Aramaic to English, see https://www.daleallenhoffman.com/

Release the echoes of my hidden past as I cancel all of my
 concerns with others
Do not let me lose my true Self in forgetfulness, but wholly release me from
 the errors of my perception
For the undivided holiness is the Absolute, the All and the Only, and the
 power of our eternal radiance from cosmic gathering to cosmic
 gathering, from age to age, from moment to moment, from now
 to now.
May these pure intentions be the rooted, fertile earth-center from which all
 my actions flow

— Ahmeyn

17

The Misunderstood Concept of Original Sin

During the time of Atlantis the energy systems of the New Humans were transmuted to a different frequency to bring them into the human kingdom, but they were still resonating closely with the members of the animal kingdom out of which they'd been developed. Consequently, there were two vital "human" factors missing from their makeup. One was the ability to reason with its accompanying gift of free will as opposed to the purely instinctive drive of the animal kingdom. The other missing factor was the concept of immortality, which can only be grasped by more enlightened and evolved members of the human kingdom. In the animal kingdom and even in the lower frequency bands of the human kingdom on Earth, the process of immortality through reincarnation of the spirit is totally automatic, with no conscious control of the process.

In the genesis parable these two factors are referred to as "The Tree of Knowledge" and "The Tree of Life," of which Adam and Eve were "forbidden to taste the fruit." In other words, at that stage they hadn't evolved far enough past the animal kingdom to be able to properly comprehend the concept of reasoning power and free will symbolized by the Tree of Knowledge of Good and Evil, or immortality symbolized by the Tree of Life.

As you will recall, the Serpent in the Garden of Eden (page 49) represented the rebel Repterrans, who tempted them with the forbidden fruit — that is, instilling in them the ability to reason along with the gift of free will, neither of which they were ready to handle at that stage in their development, being still too immature on the spiritual level to handle such abilities wisely. I do know how very jealous and protective Earth humans are of free will, which is fair enough, as it's an important component in the human evolutionary process, and right now I can hear some of you asking: "Hey, what's the problem with bestowing reasoning power and free will on a human? What harm can it do?"

The answer is — look at the horrible mess Planet Earth is in right now, which is the result of the precious gift of free will being given to beings who don't have the spiritual maturity to handle it wisely. Free will without the self-discipline that only comes with spiritual evolvement is as dangerous as a class of preschoolers being given a box of dynamite and some matches to play with.

It was this prematurely endowed gift of free will with which "Adam and Eve" were tempted that brought about the so-called Fall of Man. The results have continued to reverberate throughout the human kingdom of Earth ever since. The "fall" of course, like that of the "fallen angels," refers to the lowering of their vibrational frequencies because of wrong free-will choices being made. The DNA interference by the Repterrans certainly didn't help matters either!

These wrong choices were symbolized in the parable of the two "sons" of Adam and Eve, known in the Bible as Cain and Abel. Cain represents the lower self of humankind which overcame and killed Abel, the higher self, and this part of the story illustrates which path was chosen through the prematurely bestowed gift of free will. At that point humankind chose the lower path with its associated lower emotions and vices of greed, lust, jealousy, egotism, *etc.* all of which are born out of what the Church has dubbed "Satan," which is simply the energy of fear within human hearts.

The higher self, on the other hand, chooses the path of unconditional love, trust and oneness. In Christian terms this is Christ Consciousness, but Buddha or Krishna Consciousness are equally applicable terms. Humankind will remain trapped in a "fallen" state until such time as unconditional love is chosen over the lower emotions, which all stem from fear. It's this fear element that must be overcome to enable the conscious focus of Earth-human minds to ascend to higher frequency bands of the Human Ladder, thus it's a matter of tapping into either the "God" or "Satan" element within

The Misunderstood Concept of Original Sin

your own heart. In the end it all comes back down to taking responsibility for your own choices.

As Rumi states,

"Our task is not to seek for love, but merely to seek out and find the barriers within ourselves that we have built against it."

It was the wrong choice made back then that is behind the concept of "original sin" of which the Church speaks, but the Repterran Controllers twisted this idea to condemn half the population of Earth — the female half — as being originally to blame for the downfall of humankind. What really happened was that once the newly evolved humans, symbolized by "Adam and Eve," acquired reasoning power and free will, they began copying the questionable behavior of the fallen Repterrans. It was rather similar to the behavior of some of today's younger generation, who idolize and slavishly follow the sexual exploits, drug taking and childishly immature behavior of their favorite "pop" stars.

With this premature acquisition of free will, some of the New Humans became intractable and defiant, very much like undisciplined children, and were ejected from the "safe area" (Garden of Eden) that had been set aside for them by the off-planet Guardians. Remember this area had originally been set up to house the development program in order to keep the selected hominid species separate from other primates that were not part of the program. This then enabled further interference by way of introducing the brain parasite that was responsible for affecting their budding DNA, thus opening them to exploitation by the Repterrans.

After these New Humans were evicted from "Eden," they scattered far and wide and multiplied in number quite rapidly. The problem was that some of them began mating back with other primates, and in this way the new species began devolving back from human to animal. Some, however, did learn to use reasoning power and free will wisely, and are symbolized as "Noah and his family." They continued to evolve, and some of them are said to have had

WANDJINA CREATOR BEINGS, AUSTRALIAN CAVE

unusually long life-spans. This was because of the strong infusion of off-planet genetics in their makeup.

Those who remained in "the Garden of Eden," now known as Australia, retained a very pure and untainted understanding of their off-planet heritage, which has been faithfully passed down over generations. Their artwork (above) still accurately depicts the Wandjina Creator Beings surrounded by their aura of Light Energy. The traditional Aboriginal people still stay faithful to their role of Wisdom Keepers and Caretakers of the Land. Because their languages are very complex, with deep underlying spiritual connotations that cannot be translated into modern Earth-plane languages, so many misunderstandings have occurred. An example of this is the term "Dreaming," which doesn't only refer to an ancient past, but is really a continual and ongoing relationship with the Creator/Ancestor Beings. It is similar to how people refer to the experience of ET contact.

The event recorded in many cultures and referred to in the Bible as "Noah's Flood" was the demise of Atlantis. Noah's Ark" is symbolic of the rescue mission carried out to protect the New Humans who were progressing well. The off-planet ETs had been aware for some time of dangerous experiments involving crystal energy that the Repterrans had been carrying out, and so were ready to launch a major rescue mission when the inevitable occurred.

Crystals have the potential to be used for healing. If used properly they can help to realign the chakras — the energy grid system of the body. However, they can also be used to cause harm. Crystals do have consciousness, because consciousness is present even at the level of the mineral kingdom. They absorb and magnify energy, but they do not discriminate between negative and positive. The Repterran group in Atlantis began first using crystals as a means of reprogramming those people who were considered to be "uncooperative" or "enemies of the state." This included anybody who didn't support their bid to take over the planet.

The type of system they set up could be compared to modern-day dictatorships complete with torture techniques used on "uncooperative" citizens. Once they'd mastered the art of realigning human energy systems, they then began experimenting on the planetary energy grid. It was this experimentation that caused the demise of Atlantis. The higher off-planet Councils allowed Atlantis to be destroyed because, had the Repterran program continued, it would have hurt other off-planet energies, which could not be permitted.

As with the nuclear bomb detonation in World War II, it is another example of how the ETs will intervene with happenings on Earth — the answer is that intervention will occur if human free will becomes a threat to other races or parts of the universe. Until that happens, free will must be allowed to play out. Other races cannot generally intervene with free-will planets. It's also important to understand that the balancing of karma is usually involved, with certain pre-birth choices and decisions made by the ones involved in any event, whether that event is positive or negative. So the more of us who understand the greater reality, the more we can help shift consciousness and the more help we may receive.

The major flood and destruction event in Atlantis made a huge impact on the New Humans' collective psyche. Additionally, dire warnings were given that this "punishment" would be visited upon them again if they fell back into the "sin" of mating with members of the animal kingdom. Even the Repterrans didn't want the New Human species to devolve to that extent, because they wanted them to remain human enough to be able to follow orders and carry out tasks on behalf of the Controllers. The New Humans were led to believe that the flood was "God's punishment" for their "sins of the flesh."

Because of Repterran "editing" of the facts, the idea of the so-called original sin has become extremely confused in the collective consciousness of humankind on Earth. They know that it had something to do with the temptation in the Garden of Eden, and also that there were sexual connotations involved. Thus Eve (together with all womankind) has been made the scapegoat, as recorded in the Bible by the patriarchal Repterrans. The deeply ingrained memory of a great flood as "punishment" for this original sin — in other words, a cleansing by water, is the concept behind the rite of baptism carried out on Earth to this day.

Empower Yourself and Evolve

*"I have just three things to teach;
simplicity, patience, compassion.
These three are your greatest treasures."*

— *Lao Tzu*

18

Science and Religion

The division between science and religion that has come about on Planet Earth must be resolved if we are going to move on in Oneness. This division began with science viewing the world as a machine and mushroomed when Darwin's understanding of evolution was promoted. His theory is inaccurate in that it only shows part of the picture. He acknowledged physical evolution but dismissed the spiritual/soul aspect. The creationist story of world religions, however, is also incomplete, because it doesn't take into account natural order and progression, relying too heavily on the supernatural. The fact is that those phenomena classed as "supernatural" are simply aspects of nature that until now have been beyond the limited understanding of science.

Sasquatch Elder Kamooh: "The theory of Panspermia proposing the spreading of life and genetic codes across the Universe is right, but it also involves the participation of evolved intelligent species and spiritual consciousness."

In reality the two concepts — scientific and religious — work very compatibly together in Oneness, with spiritual/soul consciousness controlling physical evolution. It is only by gaining an understanding of both the outer physical process and the inner spiritual process that the bridge between science and religion can be rebuilt. The missing link is energy — universal life-force energy. Think of Einstein's famous theory, $E = MC^2$ (E is Energy, M is Mass and C is the speed of light). Everything in the entire universe is based on energy, vibrating at infinitely many frequencies. Evolution is a gradual speeding up, expanding and refining process through which energy passes, or "ascends."

When you think about the idea of higher intelligences working with Creator energy to manipulate life, the "Intelligent Design" concept can take on new meaning. Although most scientists think this is still religious nonsense, a few bold scientists, mostly Christian, are willing to follow the clues to see where it takes them, even if they have to rethink their own beliefs. One might think that the discovery of quantum physics over a century ago would be opening more scientists' minds, but sadly it seems to be having a tough road. There are still too many scientists who can't even embrace quantum mechanics because it feels too much like theology. It incorporates the effects of consciousness, which are unmeasurable in science labs.

Science on Earth cannot progress much further until the spiritual mystery behind human consciousness is addressed. The higher-vibrational energy behind expanded states of consciousness is the missing link between science and religion. This is why religion and the scriptures need to be understood on multidimensional levels beyond the 3-D concept of Earth-plane "reality." God is not a 3-D being, but God Energy/Essence can and does express through 3-D beings. From a multi-dimensional approach, the universe, or perhaps multiverse, is made up of energy vibrating across infinitely many frequency bands, or the electromagnetic scale. God, the source of this energy, is of the highest vibrational frequency, manifesting as Love and Light.

God/Source energy expresses through all levels, is multi-dimensional (omnipresent), and is all-knowing (omniscient). As a creation of God, the universe, too, is multidimensional, so therefore it is created in God's image. The part of this multiverse that is vibrating at energy frequencies measurable by Earth-plane science is very small compared with the whole.

Humankind exists throughout the multiverse. This is the "many mansions" of which Jesus spoke. Why would God/Source limit Its boundless presence by expressing through only one human species on one tiny planet?

There are trillions upon trillions of worlds through which God/Creator Energy expresses Itself. After all — Creator is limitless!

Science and Religion

A comment by Sasquatch Elder Kamooh on Earth-plane belief systems: "Some belief systems are suggesting that some people are inferior to you and deserve to suffer, or insisting that some souls will be saved and rewarded, while others will suffer eternally, or that only a belief in a doctrine, a magic formula or a code of laws will allow you to avoid judgment and death. Those misconceptions only keep you divided in fear, conflicts and hatred, disconnected from the Soul. Those belief systems, devised by the lower lords, keep your minds enslaved, for serving their agenda."

God/Creator Energy is the catalyst that drives evolution, and this is where science and religion need to start working together in Oneness, which can provide science (and religion) with many answers.

19

Death and Dying

Death is a major part of human consciousness that has been manipulated by the Controllers. The act of dying is a spiritual transition, and is a natural part of the evolutionary process. Evolution and reincarnation go hand-in-hand, and the death of the physical body enables us to reincarnate into a new span of existence in order to gain more physical experiences. Enlightened religions should teach about death in a constructive and positive manner, but all too often the opposite is the case. It's too often through religion that the Controllers have spread the false belief of a judgmental God figure seated on a heavenly throne, condemning sinners to an eternity of hellfire and damnation. So much fear surrounds this natural and inevitable part of life that many Earth humans don't want to think about it, prepare for it, discuss it or know about it, even though it's the only thing in life of which one can be completely certain. Such religious dogma causes a serious blockage to our natural transition back to God/Source/Oneness.

Controller disinformation and manipulation of truth to disrupt the smooth transition of the mind/spirit after death had its beginnings in cultures such as ancient Egypt. The idea of preserving the body through mummification and surrounding it with possessions "to take with it to the afterlife" was promoted. This actually had the effect of anchoring the departing spirit more firmly to the physical Earth plane and their material possessions.

Another example is the dogmatic belief in some religions that the body must be buried rather than cremated because on the "last day of judgment, God will raise all true believers from the dead." Even now in the 21st century there are still people who literally believe this concept of physical resurrection to be true! The life of

Christ is actually a blueprint of the spiritual path of human evolution. His suffering, death and resurrection are symbolic of the evolution of the spirit.

As we discussed, a person must undergo many trials and tribulations on the physical plane before physical form is eventually left behind. This enables the resurrection, or ascension of the spirit to higher energy frequencies beyond gross physicality. These concepts have been misinterpreted into distorted beliefs about the physical body being raised from the dead.

The majority of spirits go no further between physical lives than the astral plane. What is experienced as reality on the astral plane is governed entirely by a person's beliefs during their life, so the manipulation and control of human consciousness and belief is a major Controller agenda on Earth.

The Controllers have created what could be called "illusion, or amnesia nets" which trap human minds when they step free of the physical body at the time of death. In other words, if a person dies in a state of fear or guilt, fully expecting a judgmental God figure to dole out punishment, then that is exactly what the mind will create for itself in the afterlife. It will become entrapped in the "illusion net" of this belief. Conversely, if there is a firm belief that there is no afterlife and that the physical body and brain ceases to exist after death, the astral plane illusion will be an empty void of black nothingness.

The mind/spirit will then remain in this dream-state until eventually consciousness dissipates and is automatically drawn back down "onto the Wheel of Karma" and into another physical life, which is why Buddhism teaches "conscious dying," which enables the spirit to break free of this astral entrapment. It is why proper spiritual teaching and understanding need to be readily available to everyone. Our mind-state at the time of death is the controlling factor of what we experience between lives and in the next life. It also determines whether we progress to higher planes after life, or not. The Controller agenda intends to keep people trapped on the Earth plane, returning again and again with no way of escape to higher frequencies of reality unless they choose to do so. The Greater Reality is not physical, but rather is an unseen holographic energy which forms the foundation, or blueprint of what is mistakenly perceived as "reality" on Earth.

An acceptance and understanding of reincarnation as a natural path of evolution is extremely important, as it keeps the mind/consciousness moving forward and always preparing itself for improvement and growth in the next life. By removing all teachings on reincarnation from Church doctrine, and promoting fear-oriented dogma of hellfire and damnation, the Controllers have very effectively interfered with people's all-important state of mind at the point of death, thereby interfering with their spiritual evolutionary process. To repeat — *irrational fear is separation, love is oneness.* God/Source/Oneness is Love, so your only path back to Oneness is Love.

An important point also is the role played by those left behind. Some people feel that by holding onto the deceased person's ashes, they'll be able to maintain an emotional link with the loved one who has passed on. The truth of the matter is that by letting go and disposing of the physical remains completely, either through burial or cremation, a stronger and better energetic link can be maintained. Holding onto the person's ashes causes their energy body to be weakened and fragmented, thus blocking their progress. When all physical traces are disposed of, they're able to progress more easily to a higher state of being, making contact easier if it's required.

This, too, is a valuable lesson in "letting go" for those left behind. If your love is pure and true, the last thing a person would want is for the spirit of a loved one to be trapped. The preservation and displaying of religious relics such as the bones and blood of saints has the same blocking effect on the spiritual energy body, even if the deceased is considered to be a "saint" by religious authorities. I am not referring here to religious talismans such as a cross, crucifix, pentagram or Star of David. I'm purely referring to biological relics.

Most tribal people clearly understand this process of the spiritual journey after death. For instance, in the Australian Aboriginal tradition, a "wake period" is held for a short time, during which friends and family gather to reminisce fully and celebrate the person's life. After that period, the deceased is fully released to the point that even saying their name is taboo. On Australian television, if a documentary is to be broadcast that includes references to deceased Aboriginal or Torres Strait Island people, a warning is first displayed on the screen that this will happen, to give viewers from these cultures notice to change channels.

More knowledgeable cultures on Earth understand that aid and guidance need to be provided for the dying person, both prior to and after their passing. One such process is described in the *Tibetan Book of Living and Dying*, however there are other methods that can be used.

Some years ago Grey Elder Oris taught us an aura-cleansing technique to use in conjunction with Reiki. We've taught this cleanse to many people, and we have uploaded it to YouTube Channel zetaguardian1[1] to make it freely available to anyone. It's not a Reiki technique, so you don't need to be attuned to Reiki to use it.

At the end of 2012, both of Helene Kaye's parents passed over within days of each other. They'd been married for over 60 years, and everybody joked that they were "joined at the hip." At the time of her husband's death, Helene's mom began feeling pain in her hip, which turned out to be a blood clot. On the morning of his funeral she suffered a major stroke, and she passed away 11 days later, two days after her birthday.

Helene had visited them regularly, and, during their final weeks, she gave them Reiki as often as she could, including the aura cleanse. She was with them during the dying process, and continued to do the cleanse on them, making sure to do one as soon as possible after their passing.

A few months later, Helene's husband went for a psychic reading, during which her mom came through to communicate. The reader was amazed at how strong her energy connection was after

---

1. bit.ly/2iBmvK4

such a short time. Helene's mom explained that it was the aura cleanse that had assisted their transition and helped them both to move more easily through the "veil." She went on to say that others who'd passed over at the same time as they did were taking much longer to settle in and move on to where they were supposed to be. She commented on how much it had helped her on the energy level.

Many people approach death in a state of anger, fear, pain and frustration. As a result, this and the aging process begins to break down the underlying energy system of the body, which becomes imbalanced and disconnected. As the case in point described, the aura cleanse made the mother feel loved, nurtured and reconnected. It has the effect of releasing the dying person from the lower-vibrational energy of fear by reconnecting and re-aligning them to their Higher Self and the higher-frequency energy of Love and Oneness, which enables the spirit to remain free of the illusions and amnesia nets of the astral plane so it can move on to where it needs to be. It's all about energy — the higher a person's energetic frequency, the less likely they will become trapped. This understanding is the key to spiritual freedom for humanity.

Some Greys have been negatively linked to the death process through a massive amount of negative disinformation that is being spread about them at UFO conferences, contactee groups and in online radio interviews. This disinformation campaign is often carried out by Controller/Repterran "plants" who pose as speakers, researchers, "abductees," contactees and "military whistle-blowers." A classic example of this type of disinformation is that the Greys "steal people's souls." Such accusations are rampant, and in some cases photographic "evidence" is produced. These are usually cleverly faked to back up claims, supposedly of Greys in the vicinity of graveyards.

One of the Greys' main functions is soul retrieval, usually in situations where the energy frequency is too negative and heavy for higher "angelic" beings to enter. Depending on the kind of death and because of all the fear and other negative emotions surrounding death, many spirits become trapped at the time of death, unable to move properly through the veil. The Greys are far more than just ET beings — they're also interdimensional helpers who specifically work with life-force energy. (I have memories of doing this type of work in my Grey form, and I may well have been "caught on camera" carrying it out. On one occasion I helped the spirit of a young girl to escape from a public toilet building. She'd been murdered, and her spirit was trapped there.)

The other thing Greys do is assist people in emergency situations. One example is the 9/11 event. This is referenced in detail in *The Zeta Message*, and in a number of interviews I've taken the opportunity to assure families who have lost loved ones under such circumstances that no matter how badly trapped the person is or how far away they are from rescuers, nobody is ever left to die alone and unaided. There are always helpers from the other side present to comfort, guide and assist their transition.

Conscious Dying

Our Zeta lives as Zan Tu Kai and El Or Kah ended around 70 years ago, and I still retain a memory from that life of fighting to maintain my conscious awareness through my dying process — an act referred to in Buddhism as "conscious dying." As Grey Guardians we could do this, in order to carry our consciousness over after the death of the physical body. It was especially important in this situation to enable us to continue our mission with the least amount of interruption possible, even though we'd have to take on new physical forms, *i.e.*, Earth-human bodies. Conscious dying allows one to carry some or all of their consciousness out of their soul containers so we could make the necessary preparations for the next life. This is partially why you witness child prodigies in music and art.

However, once I reached the other side, I realized that I could learn more about being an Earth-human by closing off this consciousness. It occurred to me that it would be advantageous for me to experience the forgetfulness and fear, especially in connection with my ET family. I needed to fully experience the fear of ET contact to be able to understand and empathize with Earth people going through this contact. This being the case, once I incarnated into an Earth-human soul container, I largely forgot my previous life for a number of years, as most Earth humans do. Since the Level-One Earth-human brain is only able to access 10% of the potential consciousness available, even after we "wake up" to who and what we really are, we still cannot generally fully access our ET reality while operating through an Earth-human brain.

In the last life, we'd come to understand from first-hand experience how fearful and mistrustful Earth people can be, especially of anyone who looks a little different, so we clearly saw that so long as we remained dressed in our Grey soul containers, we wouldn't get very far with Earthlings. The act of conscious dying enabled us to "change our uniforms," so to speak, to take off our ET "uniforms" and make the necessary plan to replace them with Earth-human ones, while still retaining enough conscious memories to continue our mission when the time was right.

20

Reincarnation and Our Multidimensional Self

An acceptance of reincarnation assists in our understanding of the concept of Oneness as opposed to the illusion of separation many of us experience here on Earth. When we are able to recall other lives lived, both as male and female, in many different cultures and religions, it helps us to empathize with others more easily. We eventually come to understand that beneath the outer physical shell we are not so very different from each other. It also helps us to understand our connection to God, and that Creator Energy flows through all of us. Recalling lives on other planets really emphasizes this concept of universal Oneness.

Energy can neither be created nor destroyed. To put this into religious terms, God is eternal, with no beginning and no end. As an intrinsic aspect of God Energy, our soul is the same, as it is "born out of God." It, too, is immortal, and so out-lives any number of physical bodies. By reincarnating into biological forms again and again, the soul gathers many lives' worth of experience in many different cultures, which eventually teaches us to accept and respect all others who share this planet, galaxy and universe with us. Through reincarnation we get to experience life from all angles, and slowly but surely we come to realize that we "reap as we have sown," which is a biblical reference to karma (balance). "The sins of the father will be visited upon the son" is also a biblical reference to karma and reincarnation. Of course it may be taken on the more superficial 3-D level of inheriting certain traits down through the family line. However, scriptural teachings like the Bible are multi-layered/multi-dimensional and can be taken at deeper levels for deeper understanding.

This journey is also a purging and energy-refining process. Because God Energy vibrates at the highest possible frequency, if we wish to "return home to God," by raising ourselves to the highest possible vibrational frequency, we can dwell within the Oneness of Source Energy. This process is sometimes referred to as "Sanctification," which could be likened to a gem stone having to go through a long process of cleaning, abrading and polishing to reveal the shining and beautiful diamond within.

There is no way we could ever achieve this process in one single lifetime. It takes the span of many lives on a planet such as Earth, as well as lives on other worlds. We can help the process through spiritual practice, good works and healthy living on all levels of our being, which have the effect of leavening our vibrational frequency to a higher and more refined frequency band. This process is called "Ascension."

The Bible tells us to practice good works (Faith without works is dead – James 2:17) and to love our neighbor. It may say in Ephesians 2:8-9 that our deliverance is by grace only, but that doesn't mean we can just sit back and get away with negative acts of free will. We still need to take some responsibility for our own ascension process by modeling our lives upon the life of Christ and other highly evolved teachers who demonstrate unconditional love and compassion. This doesn't only apply to Christians and the attaining of Christ Consciousness. Buddha Consciousness, Krishna Consciousness, *etc.* are just as relevant, because all of these different paths lead in the same direction back to Oneness/Source/God.

In today's global society, we have the opportunity to experience varied cultures, so people have the choice of embracing the true meaning of unconditional love and compassion more readily than those who live in more isolation. To accept, respect and be non-judgmental towards our fellow humans is a major part of our journey home to God. After all, as it is stated in the Bible — Judge not or you yourselves will be judged!

Accepting the concept of reincarnation is also a key to understanding how the Repterran Controllers are able to still be here, interacting with Earth-plane humanity. Because of their ongoing issues of ego, control and exploitation of others, they too are trapped in Earth-human form, unable to evolve to higher realities.

Our Multidimensional Self

Most people on Earth understand that we are much more than just a physical, biological body. The real us is an immortal, eternal spiritual/soul essence that is independent of the biological body. The many out-of-body (OOBE) experiences, near-death experiences (NDE) and past-life memories that have been reported by countless people from all cultures on the planet support this theory. Our consciousness survives the death of the physical body and does not die with the body. It may "blank out" for a while, hence the experience of "blackness" described by some people during an NDE, but this perception is only a temporary state created by the mind.

Another example of a consciousness that could create a void of blackness or "hell" for itself can arise from a life that was influenced by a very materialistic mind-set with no belief in an after-life or God. This state would then continue until the consciousness reaches the realization that it still *is* — it still has awareness, it is still conscious, it can still think, and it realizes that it's in a void. Then panic often sets in, enabling the "awakening" process to begin. At this point spiritual helpers approach the being to assist the spirit in determining where it needs to go.

The Ascension Process

The process of our personal energy/chakra system evolving to a high enough frequency level to meld back into God/Source Energy can be referred to as "Ascension." For Christians, the life of Christ is a blueprint of this ongoing process. Being of a much higher vibrational frequency, Jesus was able to condense the process into a single lifetime. One could say that the life of Christ is an allegory and a blueprint of human evolution. The bottom line of what we're attempting to do is to raise our energy to higher and higher frequencies. Love raises our frequency and fear lowers it. This is why the

Controllers continually brainwash humanity with their ongoing fear campaign. They keep people trapped down here, unable to evolve to higher energy frequencies. Becoming aware, or "waking up" as the ETs often say to contactees, and thus breaking through this fear barrier is the key to Ascension.

Fear is harmful, love is harmless. It expresses through oneness, acceptance, tolerance and respect for others, regardless of their viewpoint, beliefs and culture. Our return home to God/Source must embrace a return to our true state of Oneness with all life.

Acknowledging this understanding and learning to think in terms of Oneness and acceptance rather than in terms of separation and divisiveness is the key to a quicker and easier journey Home. God created everything, and everything comes from God. God Energy is much greater than the 3-D Earth-human limits that we tend to impose upon It! No matter what we, as mere humans think and feel, God is the Source Energy of the universe and all dimensions therein, and not just of our tiny little planet that our limited 3-D senses are able to perceive.

21

Being Free to Evolve

"If you want to awaken all of humanity, then awaken all of yourself; if you want to eliminate the suffering in the world, then eliminate all that is dark and negative in yourself. Truly, the greatest gift you have to give is that of your own self-transformation" —Lao Tzu

The first thing to understand is what is stopping or blocking your progress. The Repterran Controllers are using many avenues both on the physical and astral planes to stop Earth humans from moving forward spiritually, but it's just a matter of waking up and keeping yourself educated so you don't get sucked in—it's that easy! Well maybe it's not so easy, but you will be amazed at how quickly your life can shift after becoming aware of what may be controlling you and keeping you stuck in fear. This might take quite a bit of self-analysis and self-discipline, but it's worth it in the end.

Sasquatch Elder Kamooh's advice is to be aware of the forces you are facing in your long spiritual evolution. "The *only* way to face them and to provide a chance for evolution to continue on Earth is by opening your Soul to higher levels of consciousness, learning empathy, compassion, and spiritual understanding. Knowing that this body you incarnate into is just a temporary vehicle for your Soul and for the collective consciousness to evolve [in], takes away the fear of dying and gives its full purpose and meaning to life."

Once you can view your life from a higher perspective of knowing that you are greater than the physical, you gain a more accurate perspective. Religions have been one of the pathways on Earth to find that higher outlook.

Religion on Earth

Most religions on Earth are valuable and help many millions of people, because they are based on love, and they teach us to support each other and our fellow humans in times of need. Unfortunately, there are still too many churches that are fear based and preach disempowering dogmas like hell-fire and eternal damnation, and a punishing patriarchal, controlling God-figure who demands flattery and praise. There are also too many churches that promote their religion over others, otherwise known as the "My God is better than your God" syndrome, which is divisive. Just ask yourself: how does it feel to be loving, kind and tolerant versus bigoted, judgmental or self-righteous? The latter behaviors don't serve anyone, including ourselves. Love is healing to all — including ourselves.

If we can get to an understanding of where each one of us is, and accept that each one of us is drawn to a particular religion, or not, we would all find more peace. Every living creature throughout the cosmos has the Spark of God within and should be respected as such! It doesn't matter which path "home" you take, or what label you make use of to call upon your God — it is all One. "Your" God is no better or worse than anyone else's God. It's all exactly the same Universal Source Energy with different Earth-human-endowed labels. To assume otherwise is simply egotistical. *And never forget – you are not a poor, helpless sinful little being. This is a Controller myth! The God Essence exists within each and every one of us, and in all beings throughout the universe, so Love is our power!*

Difference between Religion and Spirituality

Religion was initially introduced on Planet Earth to provide clear and firm guidelines to assist the young and newly created Earth-human species in developing in the right direction, which is why religions have specific rules or commandments. These rules are designed to teach self-discipline and are often culturally, rather than spiritually, based. These rules were good, and they still are for many Earth humans. Unfortunately, religion has been misused and badly exploited over the millennia by the Repterran Controllers.

As stated, a religion can be a belief system based on dogma, with an intermediary priesthood and set rituals. Spirituality is a more personalized state of consciousness. The two *can* go together, but not necessarily. A person *can* be religious without being spiritual, and a person *can* be spiritual without being religious. Plus, a person *can* be both religious and spiritual. Religion doesn't always assist the human evolutionary process, but it can. In some cases it can even block it. However, spirituality is a more direct route to human evolution.

> *Spirituality is simply about trying to the best of your ability to practice unconditional love and do unto others as you would have them do unto you.*

Despite what some think, spirituality is not about being psychic or clairvoyant; it is not about reciting prayers by rote or robotically performing rituals, however sometimes when done consciously, one can experience great spirituality. And it is not necessarily about being a "healer" or a "seer" or being able to do astral travel (which everybody does every night during the sleep state).

Spirituality is simply about trying to the best of your ability to practice unconditional love and to do unto others as you would have them do unto you, which is found in the core of many different religions. And this applies throughout the cosmos — it is not just for humans of Earth. Spirituality is about acknowledging the God Essence within, and so honoring and respecting yourself, and acknowledging, honoring and respecting this self-same God Essence within all others, no matter what culture or belief system they happen to be following in the present life.

It's also advisable in balancing your body to carry out some form of spiritual practice in order to raise the vibrational frequency of your energy system to the highest possible level. For some people this may be doing meditation, Tai Chi, Reiki, yoga, *etc.*, or traditional religious practices carried out with non-judgmental love. As many teachers have explained, including our Grey ones, prayer is talking to God, while meditation is listening to God. It can also simply be getting outside into nature and taking in the beauty of the surroundings, or listening to uplifting, harmonious and beautiful music.

SCIENCE ON EARTH

Although science may think it has all the answers, they do not. The laws of physics as currently understood on Earth do not necessarily apply throughout the rest of the multiverses, or even everywhere in this universe. And although science has progressed us forward in many ways, it is now holding us back because it has been stuck within materialism for the past 300 years. Even as quantum physicists open new doors of "hard to understand" realms, most scientists are afraid to walk through.

Until Earth scientists and researchers get past their egos and open their minds enough to come to the understanding that, as Shakespeare said in Hamlet, "There are more things in heaven and earth... than are dreamt of in your philosophy..." and that not everything can be measured by limited Earth-plane technology, they will remain stuck. Understandably, this is difficult since many who have tried to forge new ground were often shunned, lost their jobs, and ruined their careers. Most societies do not like change, but change always comes and scientists can either lead the way or get out of the way.

One of the most difficult fields of research is studying "UFOs" (Unidentified Flying Objects) or "UAPs" (Unidentified Arial Phenomena) and off-planet life as well. It is very easy to put any serious scientist into the "crazy box." But to many in the universe, UFOs are only *unidentified* down here. For thousands of off-planet cultures out there in the vastness of the cosmos who have mastered space travel, these vehicles are not unidentified. There are many different types and technologies involved.

Luckily, some researchers and scientists are ready and willing to *look outside the box* to explore other possibilities of the universe and human consciousness. One who boldly forged new ground was Carl

Jung, who analyzed four psychological human functions — sensing, feeling, thinking and intuiting. He described "sensing" and "intuiting" as processes by which we gather information and "feeling" and "thinking" as processes by which we evaluate information. These human aspects are valuable in discerning what is helpful or hurtful, which is a vital aspect for us in becoming aware of how we can, and are, being controlled.

How They Operate

The Repterran Controllers have numerous ways of dumbing down the human mind and interfering with people's well-being on subtle and not-so-subtle levels. The emphasis on materialism, consumerism, and fashion keep people preoccupied on needing *things*. The pull of the entertainment, gambling and video-gaming industries that focus on violence, sex and fear-mongering can all be disempowering. The promotion of unhealthy food and lifestyle choices serves to make people ill and weaker. Within the film industry, movie scripts can influence people both positively and negatively. Many of these war and "negative-alien invasion" movies are aimed at children and teens, starting the brainwashing campaign on vulnerable young minds. However, one of the biggest disrupters of society is through the alcohol and legal and illegal drugs industry. The overuse of these substances is effectively causing generations of humans to devolve through becoming trapped in their addictions. Be aware of how you feel around each one of these methods of interfering with your well-being. How do they make you feel?

> *Be aware of how you feel around each one of these ways of interfering with your well-being on subtle and not-so-subtle levels.*

Another is the music industry with many modern songs that imbalance and disharmonize people's energy systems through noise, discord and negative lyrics. International research by university academics highlights the soaring number of hit songs with alcohol-related lyrics. The former Australian Medical Association President, Dr. Steve Hambleton, has slammed the unprecedented levels of alcohol "glamorizing," as new research shows as many as one in five hit songs have drinking references. The number has doubled since 2001. The pattern is consistent in the U.S., with many of the hits on the pop and country charts.

Dr. Hambleton stresses, "It's not just teenagers who plug into the hit songs. Primary-school children love music, too, and even if parents are vigilant about what the kids listen to, music with bad language, talk of alcohol, drugs and sex comes pumping out of loud speakers at retail outlets or other public places."

It is important to be very selective of the sort of music you expose yourself to, because outside stimuli affect your energy. The energy system (chakras) of the body is very sensitive to the vibes of music, and a lot of modern "pop" music, heavy metal for example, is Controller-inspired to lower people's energy frequency. The underlying beat or rhythm of such music is specifically arranged to arouse feelings of aggression within the human heart, as are the questionable lyrics of such songs.

This devolvement can be seen in declining respect for self and others, a growing lack of compassion and empathy, and a decline in manners and certain etiquettes that set the human kingdom apart from the animal kingdom on Earth. In the case of some humans, the borderline between the two is becoming increasingly blurred as they drop back down the evolutionary Ladder, and, in fact, some animals exhibit a higher level of spirituality than do some humans, particularly when it comes to unconditional love!

PSYCHEDELIC DRUGS — A GATEWAY?

I'd like to add a note here in regard to trance-inducing plants and drugs such as Ayahuasca, "magic mushrooms," peyote, and created ones such as LSD, *etc.*, that are used in shamanism, because there are important points to be aware of. Firstly, shamanism, particularly of the "New Age" variety, is confined to the astral plane and no higher, which is why the energies involved are so polarized — *good* spirits and *bad* spirits — in other words, higher and lower astral planes. The whole process, like all astral-plane experiences, is human-mind-oriented, not always of a particularly high frequency and therefore potentially risky. Genuine tribal shamans express horror at the lack of training and proper understanding of so many of these New Age "shamans." In tribal systems, shamanic lore is passed down over many generations and preparations are extensive. It is not taught in a series of weekend workshops.

The risk involved is why dabbling and experimentation in this area can be dangerous without the supervision of somebody who is trained in this field. The astral is the plane of dreams, illusion and emotion, so shamanism is largely a human-mind experience, with the associated drugs assisting and enhancing the process of opening doorways of the mind to alternative realities before the "whole system" is prepared. The process may seem very "spiritual" and full of amazing revelations and magic, but it contains the potential for a lot of negative astral-plane trickery as well.

Secondly, authentic shamans understand the vital importance of thorough cleansing on all levels before carrying out their practices. They're very aware of the dangers involved, and, unlike modern-day drug users, they know to take maximum precautions in the way of elaborate cleansing rituals to protect themselves. Where mind and energy is involved, like attracts like, so cleanliness of body, mind and spirit is essential if one wishes to venture into astral-plane experimentation with any degree of safety.

21ST-CENTURY TECHNOLOGY

Many young people on Planet Earth are losing their ability to focus, as shallow distractions in the way of network socializing take over their lives more and more.

One of the main problems on Earth is exposure to technology beyond one's spiritual maturity to be able to handle it wisely, exactly as occurred back in "Eden" when humans were prematurely endowed with free will. Along with advanced technology comes a glut of information delivered at a speed and volume that many Earth-human minds just cannot handle, hence many people seem to be spinning out of control in their lives.

The other problem caused by prematurely endowed technology is lack of focus. Many young people on Planet Earth are losing their ability to focus, as shallow distractions in the way of network socializing take over their lives more and more. This is not evolvement but rather devolvement, because with a more expanded level of conscious awareness comes deeper and more concentrated mind focus, which is how higher ET cultures function.

For example, the key to operating an ET ship is total focus and concentration, with absolutely no distractions of any sort being allowed to impinge upon one's mind. In fact, it's rather similar to doing Reiki, Tai Chi or yoga in that it's a *zen* experience involving body, mind and spirit working together in complete harmony and balance, only it's much harder! I know, because I have conscious memories of doing it.

I also have a conscious memory of performing a type of Qi Gong practice as part of the preparation for piloting a ship. The commander of the ship guides all crew members through this as part of a pre-flight preparation. This process balances, harmonizes and focuses everybody's energy systems into a state of Oneness. We then connect into the energy system of the ship by placing our hands on the control panel.

This balancing, harmonizing and focusing aspect is why meditation is so important on a Level-One planet like Earth. It's a way of training the mind and controlling the thought processes into a highly focused, and therefore potent energy pattern like a laser beam. Many Earth children can barely manage to sit quietly and concentrate their minds for five minutes!

> *The ETs could teach humans much in the way of handling technology wisely.*

The ETs could teach humans much in the way of handling technology wisely. Despite incredibly advanced technology, Zeta people live extremely simple and uncomplicated lives. Many Zetas take pride in the art of *scribing*, which involves preparing hand-written documents, manuscripts, *etc*. This process is similar to the writing and illuminating of manuscripts carried out by monks centuries ago, and it is also similar to Chinese calligraphy. Scribing is done so as to maintain balance and act as a foil to the technology. Zeta family life, and what is referred to as "soul socializing" (having deep and meaningful conversation with others) is also considered important, and their life-style is very Zenlike. Therefore, it is very beneficial to keep our lives as simple and uncluttered as we possibly can!

Sasquatch Elder Kamooh speaks of their simplicity even though they are telepathic and multidimensional: "My people, like the animals, have kept this universal ability to communicate through the soul, which most of your people have lost, partially or completely. We have not developed complex grammar or syntax rules, nor a large diversity of vocabulary or spoken languages, nor devised anything of technological complexity, like your people have in their ever-changing creative evolution. But we have kept our natural ability to hear the souls of all living beings and of our home planet, Mother Earth."

Everybody in the whole cosmos instinctively seeks to reconnect with Source Energy. Older and more evolved planetary cultures are consciously aware that this Energy dwells within each of us, but many humans on Earth have not made this conscious connection, so they seek it outside of themselves.

All addictions in Earth-human society stem from a misguided search for this Divine cosmic Essence, but most of us don't bother looking inside to find it. The Controllers know this, and so, by using subtle psychological trickery, they have set up vast industries to lure people into these addictions, and once ensnared, the "victim" is very effectively trapped.

If Earth humans ever want to be free to take back their power and evolve to higher levels, then self-empowerment is a prerequisite. While people remain stuck in their own dramas and "victim mentality," that is how they will remain—helpless victims and prey to the Controllers. We are not helpless! We are immortal spirits and an intrinsic part of God.

We do have the power of Love/God within, and it is the work of the Greys and other ETs, who are here to push us until we acknowledge this fully within ourselves. This is tough love, but it must be done if we are ever to mature as evolved spiritual beings and cosmic citizens. Most people who have interacted with the "true" Greys will speak of their transformation from fear to love. And although it wasn't easy, most wouldn't change it. If the whole truth be known, it's most likely a part of our own soul essence that is the main one carrying out this plan, which jolts us out of our complacency!

Remember the Human Ladder. In the greater reality there is a part of our being on each of the ten levels of the Human Ladder. These are various aspects of our soul, or true self, which are like facets on a diamond. As our consciousness expands and evolves upwards through each of the ten levels, we become more and more consciously aware and open to these other higher aspects of Self, which is what true Self Realization is all about.

> *We are not helpless, weak sinners!*
> *We are not victims!*
> *We are immortal spirits and an intrinsic part of God!*

22

Getting Through the Fear Barrier

"If we're growing, we're always going to be out of our comfort zone."
—John Maxwell

Col. Philip Corso, of the research and development team in the U.S. military that was involved with back-engineering material taken from the Roswell crash site, had some contact with ETs himself. On one occasion he asked one of them: "What can you offer us?" Their answer: "A new world, if you can take it."

The biggest thing holding us back from that new world is fear. This seems to be a message that many contactees receive, and I want to share the wisdom I've received and synthesized from our Elders:

Fear barriers that limit human evolution/consciousness expansion may not necessarily be only from one's present life. They may very well have been brought over from many past lives, having gotten buried on very deep levels of the psyche. If Earth-human researchers really want to understand human evolution and ET contact, then a belief in reincarnation is an absolute prerequisite. A non-acceptance of reincarnation is a major block to further understanding and spiritual growth, and, in itself, it is a product of disinformation and fear.

These fear barriers that we are referring to are not so much phobias or protective precautions like a fear of heights, drowning or snakes, which are normal and meant to be protective. The fears that need to be

> *The very core of universal truth is that we are all One.*

addressed and released to enable further evolution to take place are more about negative traits that involve "the other," like racism, religious intolerance, bigotry, greed, jealousy, envy, hatred, egotism, judgmentalism, and self-righteousness. These traits are destructive to everyone. These aspects of fear are also the direct opposite of the Oneness of unconditional love, and are left over from the herd instinct of the animal kingdom by which anyone who is "different" and not "part of the herd" is at best shunned and not trusted and at worst tortured or put to death. It is this type of fear that motivates bullying behavior in schools, work places and family situations, and it is the catalyst that drives war. Until such fear-oriented and xenophobic behavior patterns are cleansed from your psyche, you cannot evolve spiritually.

It cannot be emphasized enough that the very core of universal truth is that we are all One. No matter our nation, our religion or our planet, we are all part of Oneness/Creator/Source. And to fully embrace and express this core truth of Oneness, all fear-oriented mind-sets of "separateness," "superiority," "self-righteousness," "them versus us," and "my God or religion is better than your God or religion" must be cleared away.

This divisive type of fear is like a deep-seated infection that needs to be brought to the surface to be purged and healed. Hopefully, we are witnessing this happening with the global upheaval and revelations that are shocking the planet today with leaders across the planet promoting such behavior, and the majority of citizens being appalled by it.

The following guidance is a direct message from the Elders:

My dear ones, as the Age of Aquarius illuminates the horizon of human consciousness on Earth, many more of us — the ones you call ETs — are making our presence felt among you. Some who have experienced encounters with us are often terrified by our appearance, and, because we are not seen as beautiful or attractive by Earth standards, we are often labeled as evil. Those who see us so have yet to learn that physical form is nothing but an illusion — an outer mask— and that there are many others throughout the myriad planets and dimensions of the universe who are quite different from yourselves in appearance but who also aspire to Godliness.

Some of your fear also stems from the fact that we come to assist in the birthing process of your consciousness into the new age. This process involves freeing your minds from the limitations of Earth-bound three-dimensional reality to enable you to access the more expanded awareness of cosmic reality. But, as you know, the act of giving birth is not always easy, and it is not a task to be undertaken alone and unaided.

> *In order for mind/consciousness to expand in this way, the barriers and limitations of fear must first be broken down.*

In order for mind/consciousness to expand in this way, the barriers and limitations of fear must first be broken down. This process involves bringing fears from the past and present up to the conscious level of your mind so they can be faced full-on, dealt with and purged from your psyche. These fears may be buried in the very deepest, darkest recesses of your subconscious mind, and may therefore be issues of which you are not consciously aware, but they are there nonetheless, holding you back, limiting and blocking further spiritual growth.

The vibrational frequency of your home world is being raised and refined, so your energetic frequencies must also be raised and refined accordingly. If this does not take place, the continuity of your existence here will not be possible. Even now the life-spans of some are being cut short because of an incompatibility with the changing planetary energies.

Our appointed task is to assist you past the limitations imposed upon your minds by fear. This is why many of you have issues with the Grey eyes, which act as mirrors to the Earth-human subconsciousness. They reflect back to you your own deepest, darkest insecurities, which are drawn to the surface in order to help you to acknowledge and face them. As hard and unpleasant as this may seem, it is the way out for you. This process is never done without the permission of the person's higher self. Once you have broken through the fear barrier, the "Satan" within, you will find nothing but love on the other side—and love is but another name for God.

As Planet Earth ascends and expands into Fifth-World consciousness, there are still so many among you who are ill-prepared and unready. Some, in their search for truth, have been led astray down dark pathways of fear and confusion. They are the ones in greatest need at this time, but they are also the ones most likely to fall victim to false "masters" and "gurus" on your planet who, driven by ego and greed, rise up to take advantage of the neediness of younger souls, playing upon their fears, insecurities and superstitions to control and manipulate their minds.

A truly spiritual teacher does not preach fear, hatred and judgment. Their words are simple and their only message is pure unconditional love and respect for all. Unconditional love is expressed through compassion, patience and peace. It embraces all cultures, all languages and all religions with respect, tolerance and non-judgment, never ever elevating one over another or putting others down.

The way of inner peace, simplicity and quietude for all is the only path to a deeper spiritual understanding of Oneness. At-oneness with self needs to be mastered before unconditional love can truly be extended to others. If a human heart is not at peace within itself, then pure love cannot flow forth. By pure love we mean love that is not tainted by neediness, possessiveness, ego or jealousy. This, my children, is most important for you to understand—it all begins with self.

> *The residents of Earth must learn to focus more upon similarities rather than upon differences between themselves and others.*

The residents of Earth must learn to focus more upon similarities rather than upon differences between themselves and others. You may think that we look very strange, frightening and "alien," but truly we are not so different from you.

OVERCOMING FEAR

I'll close this chapter with a teaching that was given to Helene Kaye by her Elder Oris, which hopefully will assist you in overcoming your own fears—the fears that disempower you.

Fear can teach you about the unknown, and push you through the veil of self-doubt. Some veils are thick, and some are thinner, depending upon your experiences from past lives and from your childhood in the present life. There are many fears and many lessons to learn.

The biggest lesson of all is to know, feel and understand that Love is your power, your strength and your guidance. Think of your fear as the thick, blurry lenses of a pair of "invisible glasses." Through learning experiences, forgiveness and deeper understanding, your lenses become thinner, until eventually your "glasses" disappear. Stop judging yourself, and allow for bumps in the road. They are there to slow you down, to enable you to release the pain and to give you the time and space to find out who you really are. Consider your "glasses" also act as a mirror:

Do you attract positive or negative experiences into your life?

Do you attract positive or negative thoughts into your mind?

How much do you love and respect yourself?

People who have huge ego or anger (fear) problems do not feel loved or good enough, and "mirror" their own insecurities. Remember, we attract to us what we need to learn or heal within self, on multidimensional levels.

Your soul spans many dimensions and has many connections.

The time is now to step out of your comfort zone and push through your fear barrier. Sometimes it may feel like you're wading through mud, but when you reach the clear, crystal waters of love, self-empowerment and connection, a whole new and magical world will open up for you.

Even negative experiences with others are lessons for your spiritual growth. Such experiences are often created out of lack of knowledge, or understanding and misinformation. Sometimes you put on another person's blurry-lensed "glasses" and take on their fear as your own. Fear feeds off fear and spreads far and wide to catch you in its net. If you could only realize that the "fear net" is of a lower vibrational frequency and that you have the power to walk straight through it unscathed by trusting yourself and your own inner voice. You need to understand clearly that your own Love Energy is a much higher frequency, and is unlimited.

Out of respect and love for yourself, ask your spiritual family for help and guidance. This will help you to understand that the Love Power is a connection to self, to All and to Oneness. That is why we call it "Source." Once you learn to tap into Source, your vision will be clear, your heart will be filled with love, and your mind will be free to create endless possibilities.

Here is a list of common fears, and the antidote to counteract them. These could be used as a mantras:

I fear not being good enough — I am worthy

I fear not being accepted — I am empowered

I fear being laughed at — I am happy

I fear success — I am successful

I fear failure — I am learning and evolving

I fear being alone — I am one with all

I fear not being loved — I am love

I fear lack — I am abundance

I fear trusting others — I trust in me and my inner wisdom

I fear not having time — Everything happens in its own perfect timing

I fear negative beings — Love is my power

I fear relationships and friendships — I allow positive, loving people into my life

I fear my true potential — I AM

I'll close this chapter with two more perspectives that concisely sum up what is happening on Earth right now and in the future.

The first message is taken from a conversation between Alcheringa and Valerie Barrow from her book *Alcheringa*.[1]

Alcheringa: "There is a quickening. Part of it is affecting time — then there is an awakening of love, and those who follow it will enter into an age of peace and harmony."

Valerie: "And those who don't?"

Alcheringa: "They won't be able to survive the higher vibration, the speeding-up part. It's okay, they'll just incarnate on some other Earth-like planet with a lower vibration and continue the journey. Everybody succeeds eventually, the power of love conquers everything in its path."

And the following beautiful prophecy given to us from the Hopi Elders on June 8, 2000:

You have been telling people that this is the Eleventh Hour, now you must go back and tell the people that this is the Hour. And there are things to be considered...

Where are you living?
What are you doing?
What are your relationships?
Are you in right relation?
Where is your water?
Know your garden.
It is time to speak your truth.
Create your community.
Be good to each other.
And do not look outside yourself for your leader.

Then he clasped his hands together, smiled, and said, "This could be a good time! There is a river flowing now very fast. It is so great and swift that there are those who will be afraid. They will try

---

1. Barrow, Valerie, *Alcheringa*, pp. 156-158

to hold on to the shore. They will feel they are being torn apart and will suffer greatly. Know the river has its destination. The elders say we must let go of the shore, push off into the middle of the river, keep our eyes open, and our heads above the water.

And I say, see who is in there with you and celebrate. At this time in history, we are to take nothing personally, least of all ourselves. For the moment that we do, our spiritual growth and journey come to a halt.

The time of the one wolf is over. Gather yourselves! Banish the word *struggle* from your attitude and your vocabulary. All that we do now must be done in a sacred manner and in celebration.

We are the ones we've been waiting for.[1]

---

1. http://www.awakin.org/read/view.php?tid=702

Our Future

"I alone cannot change the world,

but I can cast a stone across the waters

to create many ripples."

— *Mother Teresa*

"

23

Universal Spirituality

Our Guardian Elders also want to share their wisdom and perspective on spirituality.

We of the Star Nations also feel affinity and accord with the various spiritual teachers who have visited your planet in the past — ones such as Jesus, Buddha, Krishna and many others. Our ways too are of love, gentleness and peace, which are virtues that all of these teachers have tried to impart to humanity, for it is only by following the path of love and peace that you will ever attain reunion with The One you call God.

There are many spiritual paths that may be followed, and it is important that you find one to suit your own individual needs, but remember, my children, do not become so distracted by the scenery along the way that you forget your ultimate destination.

Some souls in their quest for God become distracted completely by ritual, dogma, "miracles" and other phenomena of the astral kind. Instead of taking the time to nurture the spiritual virtues of love, patience, tolerance and compassion, and quietly listening to the still, small voice of God within, they prefer to lose themselves in ritual, dogma, superstition and astral plane trickery.

It is so easy to forget God's place in your life. True spirituality is beautiful in its simplicity, and there is really no need for ritual, dogma or great psychic ability when you travel the spiritual path. All that is truly needed is love, for where there is love God is present also, for God/Source is love. The path you travel may be likened to a circle, with God at the beginning and at the end, so no matter what you see, or do, or experience along the way, do not ever lose sight of this, for God is always there, ever-present in everything you undertake upon your life's journey.

Our Elders have shared the 11 Universal Laws and 11 Spiritual Laws of the Cosmos, which are as follows:

THE 11 UNIVERSAL LAWS
1. Free Will
2. Change
3. Movement and Balance
4. Innocence, Truth and Family
5. Symmetry
6. Life
7. Light, Sound and Vibration
8. Judgment
9. Nature
10. Love
11. Perception

THE 11 SPIRITUAL LAWS
1. Freedom of Man
2. Growth of Man
3. Strength, Health and Happiness
4. Protection of Family
5. Equality
6. Choice
7. Intuition
8. Karma
9. Protection of Man
10. Healing
11. Future Sight

These Laws apply across the entire universe, and reference to "Man" includes all humanoid-type people of all galactic cultures, not only Earth-human males.

Because of their limited 10 percent-conscious awareness, most Earth humans don't have a conscious memory of between-life choices and decisions made, but the truth is that everybody makes this pre-birth life plan, based on what lessons are required to enable maximal spiritual growth and karmic balance in the coming life. Quite often Earth-human interaction with ETs, and most especially with different-looking ones such as the Zetas and Greys, is a major learning experience chosen by Earth humans to help them understand that they are not the only human beings in the universe, and to give them a greater appreciation of the "many mansions" that make up cosmic reality. Ask any contactee. Seeing an off-planet visitor for the first time has a profound impact on one's idea of reality. This preparation is most important if the people of Earth are to someday become members of the Star Nations. From the non-attached and bigger-picture perspective of the higher self, such choices and lessons may not necessarily be what the physical, or the lower self wants, but they are always what we need.

* * * *

Many Earth humans have great difficulty with this concept, particularly when it comes to *any* unpleasant life events and circumstances. For example, how many are able to accept that they've made a pre-life free-will decision to suffer a major accident or illness in their present life, or to be born into poverty or an abusive family situation? When such things occur, the "victim of circumstances" usually blames God or bad luck. However, like it or not, we do choose freely, but this choice is made from the Higher Self "bigger picture" perspective, and there's *always* a positive learning aspect involved, either for self or for others around us.

To help illustrate this, I'll use an event that is a perfect example, but it doesn't involve Earth-human sensitivities, so nobody from down here should be upset by it. I extend my apologies to any ET readers who are affected, but I know you'll understand clearly the point I'm trying to make. Let's look at an event such as the Roswell disc crash in which two ET ships came to grief near Roswell, N.M. in 1947 because it involves both on and off-planet folks.

Viewed from the limited three-dimensional perspective, this event appears to be a tragic and bizarre accident, but at deeper levels of consciousness, it was pre-planned. To understand this, the event needs to be explained from two perspectives. On the higher level of consciousness, the off-planet ones involved in this "accident" knew exactly what was to happen and why. They knew very well that this "accident" was an ideal way to bring Earth-humans' attention to the fact that there really are other beings out there in the vastness of the cosmos; that some of these are real physical people who travel in physical craft; and that they are people — they are not God, they aren't infallible, they can make mistakes and have accidents, and in fact they are fully human. In other words, the event had within it a number of subtle messages for Earth humans to pick up on.

It was also an ideal opportunity for Earth humans to get to physically touch and examine off-planet beings and their technology in a non-threatening way, and to hopefully learn from this. Allowing the Earth authorities to interact in this way enabled them to remain empowered, which may not have been the case if the ETs had landed and approached openly. The communication gulf was just too wide. Panic would have ensued, and Earth people may have been hurt, either physically or emotionally.

On the other hand, the ETs, with a deeper understanding of their true "soul nature" and the fact that "they" are not their physical "container," would have been willing to sacrifice their physical lives to this end, knowing clearly and consciously that the real "them" — the soul consciousness — is indestructible and cannot die. As it played out, the Earth authorities were able to maintain their position of control, while still receiving clear evidence of the reality of off-planet visitation. It was this bigger-picture scenario that the ETs had planned for.

This is not to say, however, that it was easy or pleasant for those involved. On the human lower-self-level, the pilots and crew of the ships were totally shocked and terrified, as anybody would be under such circumstances. The ET crew looking on from a third ship who witnessed the event were equally affected, and none of them had set out on the mission with the conscious knowledge that they were going to crash, so from that perspective it was a tragic accident. The ones involved chose not to retain conscious awareness of what was going to happen, because if they had, they may not have been able to face going through with it when the time came. The fact that it's been so effectively hushed up down here was beyond the ETs' control, which is where Earth-human free will intervened, to block the message from the general public.

This concept of lower- and higher-self-awareness is very difficult to express in three-dimensional terms. It's like two separate time-lines running parallel with each other, and the limited lower-self-mind is not consciously aware of the broader and deeper "bigger picture" understanding of the higher-self-mind, but this doesn't mean that the broader scenario is invalid.

This example can be applied to many human events on Earth both the large ones that get everyone's attention, but even the smaller ones that affect families every day. The choice is always ours as to whether we'll "see" the bigger picture, or remain locked in fear and victimness. Gratefully, with many of the terrorist events around the world, love is prevailing.

For this reason, here is a short meditation to assist in healing for those involved in both large and small tragedies.

We wish to send healing to all who died or were injured in any horrific event. We wish to send healing to the Group Mind of those who were involved, witnessed, or felt any heartache from this event. Let this healing release any guilt from their hearts. Let this healing reconnect their hearts and souls and bring joy, harmony and connectedness to all. Let this healing bring balance back to their group.

May this healing ripple out from Earth to the Cosmos — from Earth Humans to Cosmic Humans! May this energy heal the connection between Mother Earth, her people and their cosmic family!

May we all raise and transform our vibrational frequency in love, and may we rejoice in our connection to Oneness!

Blessings, Love and Peace to All.

24

Interdimensional Emotions

One of the claims often made about the Zetas and Greys is that they don't experience emotion. This is not correct. Zetas and Greys are insectoid human beings — not mammalian human beings like Earth humans, but human beings nevertheless, and all human beings throughout the universe are capable of experiencing emotion. They do express emotion that one can pick up when around them — it's just not very visible.

Emotion is simply a form of energy vibrating at a certain frequency. Greys operate at a higher frequency than do Earth humans, so their emotions express at a higher frequency that is not physically perceptible to Earth humans. However, many sensitive humans taken on board our ships are able to sense deep emotion coming from us. I clearly and consciously recall experiencing emotion as a Grey, and when I'm there, at that level of reality, emotion feels no different from when I feel it down here as an Earth-human.

People may find this hard to believe, but Greys also tend to have a very well developed sense of humor, which is most necessary, given the type of work we/they carry out. Their humor tends towards "dark humor," as with many emergency and health-care workers on Earth. I also remember getting together for some great "parties" up on the ship, as a way of coping. Many various cultures come together at these gatherings, and the sense of love and oneness is beyond words to describe. Every little "win" we have in assisting and/or awakening somebody down here is celebrated with dance, song and other festivities.

Human Emotions

There are six emotions that all normal, healthy human beings throughout the universe are capable of experiencing. These are

- love
- sympathy
- grief
- anger
- fear and
- joy

and all six are legitimate, normal emotions that need to be expressed if one is to live a healthy, whole and grounded existence on one's home world, whether that home world is Planet Earth, or a planet in the Pleiades star system, Zeta Reticuli star system or wherever.

The two emotions that are currently being exploited and manipulated big time on Earth by the negative forces are anger and fear, and please be aware that these two can often be linked together. Fear can cause anger and vice versa. Beyond that however, these are pure and legitimate emotions that do need to be expressed at times by any human being in a healthy way. Let me repeat, healthy way. They are not "bad" — they are normal, and to try to tamp them down completely will inevitably lead to health problems. The emotions of anger and fear, when expressed cleanly and purely, and balanced by unconditional love, are perfectly legitimate and at times necessary.

Anger, when applied correctly to right a wrong or to set one's boundaries, is part of being an empowered human being. It's when we allow our anger to become *uncontrolled or exploited* by others, or when we hold on and refuse to move on, that this emotion becomes destructive and negative. It is important that anger be expressed nonviolently, and with clarity and love for the situation to be corrected, and then move on. Nothing better demonstrates a strong, passionate use of anger as that of the non-violent protests in the Civil Rights Movement that were promoted by Martin Luther King, Jr. in the 1960s. And there were results. However, if anger begins to control your life and your way of thinking, it actually becomes disempowering and debilitating for you.

Fear, as we just covered, when used sensibly to protect yourself from entering into a dangerous situation is also a necessary human emotional response. Keeping your intuitive fear mode finely tuned is necessary, because there are both non-corporeal and physical beings around us whose motives are not always of the highest intent, and in life we do come up against dangerous situations at times. Our physical body is the temple of our soul, and to unnecessarily place it in danger is wrong. It's the emotion of fear that keeps us safe and out of harm's way, providing we don't allow it to dominate our whole way of thinking.

Lest you forget, Controllers can exploit the emotions of fear and anger on Earth as a way of manipulating and disempowering people, so it's important to be aware of this. In other words, learn to think for yourself and to discriminate between pure and impure emotion. Learn to discriminate between fear that provides genuine protection and that which is manipulative or exploitative. Learn to gently but strongly set your boundaries so that you are not controlled by anger.

Some belief systems on Earth tell us that to be truly spiritual we must practice nonattachment from the lower emotions, but there's a vast difference between nonattachment and denial. Many people tend to get a bit confused over this, thinking they're practicing nonattachment but in reality are simply in denial, which is not a healthy state for a human mind. Believe it or not, one can express both anger and fear and still remain nonattached. The trick is not to get caught up and trapped in the emotion. Try to stay centered in love, then express the emotion and move on.

Speaking from the Grey perspective, they, too, can experience both anger and fear, but the higher one's conscious awareness evolves up the Human Ladder, the more balanced one's emotions become and therefore, are expressed. From human's perspective, their display of anger may appear to be very passive, but the energy behind it makes a clear statement. They just express with love and compassion.

Because Planet Earth is at Level One, everything, including emotion, is very polarized and expressed in duality — extreme highs and extreme lows, in other words. At higher levels this duality is much less pronounced.

It's also important to understand that fear is an emotion carried over from the animal kingdom, where it plays a role in survival. An evolved human, however, is consciously attuned to the energy of Source/God, and accepts the universe as a divinely unfolding plan rather than a haphazard event, so in the greater reality Churchill was right — there is nothing to fear except fear itself. When one can view life as everything is as it should be, at the perfect time, in the perfect place, fear and anger recede in prominence.

Zetas and other more highly evolved ETs experience anger when they see how Earth humans are being manipulated and exploited, and also the way the Earth is being hurt. It's this sense of anger and injustice that keeps them fired up enough to remain on Earth to try to correct the situation. Many Earth humans have no concept of how deeply they're being exploited and disempowered, and, as caretakers, it is the duty of off-planet allies to step in however they can. Depending upon how the disruption is affecting other worlds, they cannot just always stand back and allow a given situation to continue!

Another myth that is put out by some is that as we evolve spiritually our lower chakras gradually close down so that we eventually only operate through our higher chakras. This is not correct. To remain healthy and grounded as human beings in whatever planetary system we happen to be experiencing life at the present time, all seven of our main chakras must be operating in balance. The heart chakra (love center) is the bridge and balance point between upper and lower chakras. As we evolve to higher levels of spirituality, our whole chakra system begins vibrating at higher and higher frequencies of energy, but no chakras close down as a result of this.

INTERDIMENSIONAL EMOTIONS

Message from Elders

We also experience hope, joy and sadness. Despite what many of you have been led to believe, we do understand emotion, but at a different energy frequency that is not readily perceptible by Earth-human senses.

Not having the same musculature in our faces, we cannot physically express emotion as it is expressed on Earth, but this does not mean that we do not feel it. Just like you we can laugh and we can cry. Your love and acceptance is important to us, as we hope our love is important to you, for it is only through love that the long and sometimes difficult journey that all of us are undertaking can be satisfactorily completed. Love truly is the only path that leads us home.

25

Demonizing of the ET Visitors

By now you may be getting the message that if we are going to move into the future as Universal citizens, we will need to acquire a lot more acceptance of those who are different from us - first on this planet and then beyond. And it is important to understand how off-planet visitors have been around for a long time.

The lack of this ability is seen everywhere, but one claim being made by a number of fundamentalists is that the ETs are real, but that they're demons. Rev. Michael J. S. Carter, M. Div., addresses this issue in *Alien Scriptures: Extraterrestrials in the Holy Bible:*

"Times do change; we have TV shows like Ancient Aliens where these ideas (of the possibility that angels could really be extraterrestrial beings and that these beings made appearances in the Old and New Testaments and the Koran) and others regarding humanity's possible extraterrestrial origins are now routinely discussed. More evangelical writers like Timothy J. Daly, Chuck Missler, Mark Eastman and Patrick Cook weigh in with their writings on ETs. However, their acknowledgement of ET life focuses on a more sinister religious agenda, often interpreting 'end time' scenarios from the Bible. This interpretation of ETs is not new, and it continues to be pervasive. The 2000-2015 NBC show Falling Skies has our visitors being hell bent on taking over our blue-green planet and killing us off and/or enslaving humanity. Simple logic suggests that the visitors, being able to get here and make crop circles, etc., are far in advance of us, and if they wanted to kill us off, we would already be dead, but we are not.

But humans seem to like feeling scared.

"Although I do not agree with these writers about the agenda of extraterrestrial life, I do realize that it is so very human to demonize and marginalize those who are perceived as 'the Other,' especially when we do not understand them and we allow ourselves to be ruled by fear. Xenophobia is alive and well on Planet Earth regardless of whether one is a so-called human being or an Extraterrestrial Biological Entity."

And from p. 58:

"As a species, we human beings are infamous for fearing and demonizing that which we do not understand."

Rev. Carter discusses his own ET contact experiences with the Greys:

"It was my fear that was the block to really experiencing the visits and being able to deal with, instead of react to, my visitors. Fear keeps one from being in relationship with the world and with ourselves." (p. 18)

"I feel that my spiritual growth was somehow accelerated and that in spite of the initial fear it engendered, this was a powerful and positive experience for me!!!! My opinion and my experiences are not those that receive the most publicity, however. It seems that our culture's fascination with this phenomenon caters more towards the sensational and so-called negative aspects of the visits. Fear, like sex, sells big time in this culture, and thus those are the stories that get the airplay in the media." (p.22)

For a more accurate and less-biased view than that presented by the media, Hollywood and fundamentalists, please check out an extremely comprehensive survey carried out by — the Dr. Edgar Mitchell Foundation for Research into Extraterrestrial Encounters (FREE). This organization, initially founded by Reinerio Hernandez, an IRS Attorney, was incorporated in 2015. Respected researchers such as Dr. Rudy Schild (Harvard Astrophysicist), Dr. Bob Davis (Neuroscientist), and Dr. Jon Klimo (Transpersonal Psychologist) are on the Research Committee of FREE, and the late Dr. Edgar Mitchell was Patron until his death in 2016. I have taken part in this survey and can recommend it. The requirement is that your ET contact information/memory has not been accessed through hypnosis.

Extraterrestrial Presence on Earth

MADONNA WITH SAINT GIOVANNINO, BY DOMENICO GHIRLANDAIO

Phase 1 and 2 research data and methodology is available at the FREE website.[1]

I love Rev. Carter's comments on skeptics, from p. 35:

"It is mind-boggling to me that orthodox Christianity relies upon the 2,000-year-old writings of a group of individuals, concerning a story about a Jewish Rabbi who was crucified by Rome for high treason and allegedly rose from the dead, but cannot believe the modern phenomena of multiple witness sightings (by credible individuals like astronauts and military and civilian pilots), not to mention video evidence regarding the existence of UFOs. Though we have modern evidence, this is not a modern phenomenon; UFOs have been visiting Earth for centuries."

On pp. 44-5, he includes two 15th-Century paintings, *The Madonna with Saint Giovannino* (above), by Domenico Ghirlandaio, in which a man is looking up at a disc from the field, and *Annunciation with Saint Emidius* (below), by Carlo Crivelli, both of which feature UFOs overhead.

---

1. bit.ly/2hJA3qi

"Annunciation with Saint Emidius" shows a UFO between the buildings with a beam coming out of it and into the top of Mary's head

These two examples are just a few of the many more paintings and tapestries from centuries past wherein UFOs or ETs are clearly depicted. It is beyond belief that artists would conspire to imagine similar objects in the sky, and then spend may hours implementing such figments of their imaginations while residing in different locations during a period of time when people were relatively isolated from each other.

Occam's razor is a well-respected principle from philosophy. If there exist two explanations for an occurrence, the simpler one is usually better. In this case, the existence of more advanced beings among us is the simpler explanation.

26

Star Children

Besides the people who are here as blended souls in order to help humanity evolve, there are also many children being born who have already made an evolutionary leap.

OUT OF THE MOUTHS OF BABES!

The Star Children are also known as Indigo, Rainbow and Crystal Children, and probably other labels as well, because Earthlings like to divide everything up into separate little boxes with labels on them. In reality, they're all the same in that they belong to more recent generations of genetically upgraded children who have been born on Earth over the past few decades. These children are part of the ongoing assisted ascension program being carried out on Earth to help people and the planet move up to the next level of the Human Ladder and away from the Controllers' influence.

Needless to say, the Repterran Controllers are interfering with this process. Many of these children are very sensitive psychically, and so are easily manipulated to play right into the Controllers' agenda. The format most often followed is to start by remote-mind controlling a young child into an altered state, then feeding them complex scientific or mathematical formulae way beyond their age and understanding. This gets everybody's attention, usually freaks out their parents, and the child is either placed under psychiatric care or else labeled as a Star/Indigo/Crystal/Rainbow Child with enhanced DNA.

The next step in the process, providing the child hasn't been dosed up with medications to "normalize their behavior," is to introduce either "gloom and doom" predictions through the child's Repterran-controlled mind, or else warnings of "negative ETs" who are going to invade Earth. Because of the amazing level of "intelligence" displayed by the child, of course everyone then faithfully believes everything they say while in this altered state.

My intention here is not to cause fear but just to make people aware that this type of interference can potentially occur, and is why knowledge and understanding is so important as a protection for these beautiful little souls. What occurs even more, but is very destructive is how Star Children are also being exposed to negativity and violence in the way of music and video games and through genetically modified food with artificial colors and other unnatural additives, which impact upon their energies to throw them out of harmony and balance. None of these unnatural additives are good for any child. However, being so sensitive, they're particularly vulnerable to such subtle and long-term interference.

Some Controller-oriented researchers are laying the blame for the rising level of autism in children to negative interference by the Greys and their hybrid program. Admittedly, the infusion of ET genetics is involved indirectly, but it has more to do with the off-planet genetics causing these children's energy systems to vibrate at a higher frequency, thus making them more sensitive and vulnerable to the harsh and heavy energy on this planet. Life on a Level-One planet like Earth is not easy to cope with! As a star seed, I speak here from personal experience.

An extremely important requirement for children is to allow them to have "down time." Many children complain of being bored, and so some parents, believing their children need to be entertained every waking hour, are too quick to rush in with options by filling their days with activities like music lessons, sports and other extra-curricular activities. Consequently, many children cannot draw from within to tap imaginative inner resources to entertain themselves in positive ways. Psychologist Susan Stiffelman, author of *Parenting Without Power Struggles*, says that children need unstructured time, and plenty of it, if they are to nurture their creative instincts, rather than rely on someone or something to keep them engaged.

Children who lose the capacity to daydream become restless adults, always searching for distractions and quick-fix stimulation. Boredom isn't the end of the world. It is the start of new possibilities and a fertile ground for nurturing creativity and self-reliance. Daydreaming is particularly important for the Star Children, as it is their inner world that needs to be given the time and space to expand and develop. Watching endless television and being captured by countless video games and social networks does not allow the necessary space for this inner growth, which goes back to my comments in the last chapter regarding the need for human minds to open up to realities beyond the limits of 3-D perception. Many Earth humans are getting more and more removed from balance and harmony. For too many the sense of wonder and imagination has been lost, replaced by cynicism, close-minded skepticism and derision. The ability to imagine and visualize is such a major part of being a happy, healthy and spiritual person, but there are many who feel that magic, fun and laughter are a complete waste of time, or even wrong. This is part of the reason why so many on Earth are stressed and angry, and so disconnected from themselves and the greater reality.

To quote from a teaching given to me many years ago by a Grey Elder:

"The greatest veneration you can give to God is laughter and joy. Joy is spirituality — spirituality is joy, and to be truly spiritual this joyfulness must be expressed in everything you undertake in your daily life. Strengthen your connection to Source through laughter, song and dance. Experience Oneness all around you in the simple beauty of a flower, a sunset, a blade of grass, or a forest filled with morning mist. Let It close over and wrap around you like a protecting mantle, filling your whole being with peace and love."

Too many children on Earth are being forced to grow up too quickly, and as a result are losing their natural sense of wonder and magic. Some poor little ones are hardly allowed to be children any more, and many have very little laughter and joy in their lives. A major role of the star children is to bring back balance and healing to

Mother Earth, and so this all-important aspect of wonder, joy and imagination must be encouraged, protected and nourished.

In modern society imagination tends to be put down and dismissed. "Oh, it's just your imagination!" or "It's all in your mind!" are common phrases used to put down this extremely important component of human make-up. The human mind is not an enclosed room! It is your doorway to other realities! The imaginary games and imaginary friends of childhood are very often important clues to past lives and/or spirit guides in the present life.

DIFFERENCE BETWEEN STAR CHILDREN AND HYBRIDS

Star Children are essentially Earth children who have been energetically upgraded with ET genetics and enhanced DNA. Hybrids are essentially ETs who have some Earth-human genetic material in their makeup. Since quite a number of the Star Children are being interfered with by the Repterran Controllers, more hybrid children are being introduced on Earth at this time as preparation for eventual disclosure. Having a higher percentage of ET genetics in their makeup, they have more protection on the energy level to block such interference from occurring.

The Controllers are involved in this side of things as well, producing their own hybrids by using genetic material taken from the bodies of ETs who have died in crashed ships, and mixing it with reptoid genetics, which is causing major confusion.

WHY DO WE NEED TO EXPERIENCE PHYSICAL LIFE?

Sasquatch Elder Kamooh explains: "Souls must incarnate in order to develop higher consciousness through a diversity of experiences and forms, allowing them to learn the spiritual understanding of existence. Therefore, biological life is required."

Earth humans experience physical life to polish ourselves spiritually to enable evolution to take place! Sure, when a soul first sets out on its journey to experience physicality as a human on a Level-One three-dimensional planet like Earth, they come with the best intentions, but there are just so many temptations at this level, as the Draconians once found out to their detriment. At the same time, the off-planet people being "consciously" born here in Earth-human form, as I was, are also coming as part of the mission to assist Earth plane humanity past the "stagnant pool" created by the Repterran hijacking of the planet. Up until now, only a few of us have come, but now we are coming in thousands. We are not here to be put up on pedestals, to create cults or to be worshiped as gods or saviors. We are not here to incite wars against the Establishment. That is not our way! The genuine off-planet people come in quietly "through the back door," walking our talk and doing our best to set an example for others of peace, tolerance, respect for the environment and unconditional love.

Sasquatch Elder Kamooh: "In the Cosmic Order protected by the Star Council, inter-species peaceful spiritual communications and relations are encouraged as a way to accelerate the karmic healing and the dharmic spiritual evolution. Species and souls with higher levels of consciousness and spiritual wisdom are thus asked to teach the younger species and souls, and supervise their evolutionary process through space-time and aeons."

We try our best to steer clear of creating karmic ties, but it's not always easy. We can "fall down the Ladder," too, and get stuck here, having to work our slow and painful way back up again. This has happened with our people too, so we speak from experience.

Evolution up the Human Ladder is a return to Oneness, or to put it in Christian terms, a return home to God. We set out on the journey as a pure and clear soul essence freshly sprung from Source, but gradually over the span of lives living in the grossness of physical form, we build up layers and layers of dross/fear in our mind/consciousness to weigh and anchor us down, and separate us from Oneness. We become like a rough stone that has built up much dirt and sediment over time and needs then to be gradually abraded and polished back to diamond-like perfection. Looked at in this light, the Greys (and others) could be likened to the master gem cutters and polishers of the universe, smoothing our rough edges just so we can climb back up the Ladder to Oneness.

Some people ask, "why do we ever leave the comfort of Creation in the first place?" A really good analogy for this concept is the story of Shakyamuni Buddha, AKA Prince Siddhartha Gautama, who was born the son of the King of the Sakya clan. At the age of 29, Siddhartha Gautama gave up his comfortable and sheltered life at his father's palace and went out into the world in search of the true nature of all existence. He studied with great Hindu teachers and engaged in many austere practices, almost dying from malnutrition. These experiences led him to the understanding of the need to follow the middle way, between the extremes of the over-indulgence of his palace life and the self-mortification of his more recent life.

People down here get so caught up in the need for experience that they hate the thought of death. They dream of immortality in physical form. A strange thing I've found is that so many who believe in God, including ones such as nuns and priests, are so scared of the concept of death! Earth humans are rather weird that way!

27

Understanding the Grey Guardians

"Knowledge is power; understanding is empowerment."
— *Maris*

On Unity & Grey Consciousness

Some researchers on Earth claim that the Greys operate through a "hive mentality," with no free will, but a better description of their consciousness would be "collective." I know this, because I have clear conscious recall of life as a Grey working up on the ship. It's true that we don't operate through free will, because once an entity reaches higher levels of the Human Ladder, they begin operating through God's Will. At this level one thinks more in terms of Oneness, so people are much more homogenized and amalgamated as a Group Soul Consciousness. Energy is always directed for the good of all, whereas Earth humans tend to be segregated, competitive and divided, and more self-centered, hence all the racial and religious tension that goes on down here.

In Joan Ocean's book *Dolphins Into the Future*, she cites a message and some advice given to her by her dolphin friends on what she calls Higher Consciousness. This is exactly how the Greys operate as a group consciousness. To quote:

"We merge our energies with each other, but none of us lose ourselves in this merging; we are still individual beings who maintain our autonomy. Remember to hold your space in the midst of chaos and community, in all partnerships and relationships. Do not lose the integrity of who you are and what you represent. Stay true to your chosen path. Be aware and do not compromise yourself."

Understanding the Grey Guardians

From the Elders...

Within the Zeta Soul Group we are one — one Zeta is all Zeta. Every act we perform is for the good of all, and because we are highly telepathic, generally communicating by thought, there is no dishonesty or deceit. Neither do we look upon others as being separate from self — we are all One.

Those of us who dwell on higher levels of reality feel great sadness to see the depth of prejudice, intolerance and incredible blindness of many Earth humans. The color of a person's skin, the language they speak or their physical features are so very unimportant. It is the depth of love and compassion in the heart that counts above everything else. Looking from above we hardly even see the physical characteristics of a person; all we perceive is your light and energy flow, and it is the brightness and clarity of your light and the harmonious flow of your energy body that truly matters — certainly not your physical appearance!

We will continue to stress how important it is to understand that the race, creed or culture that a person belongs to is simply a role chosen for the span of one single life-time, just like an actor taking part in a play. You would be very childish and immature indeed if you believed that the actor really *was* the character that he was playing, wouldn't you! Of course he has to learn his lines, prepare and study for the role and immerse himself fully in the character he is to portray, but after the day's rehearsal or the evening's performance is over, he takes off his costume and makeup and goes back to being himself again.

Your lives on the physical plane are exactly like the characters in a play. Before you reincarnate you must prepare for the role you are going to play in the coming life. You study the planet, country and culture that you are going to be born into, and also your prospective parents and family — in other words, the role you are going to play in the coming life, but all the time you know that it is only a role — a part in the Play of Life — and you are only acting.

When you die and return to spirit it is just like the actor removing his stage makeup and costume after the performance and going home. Once you are back in spirit you can relax and be yourself again, without the theatrical costume of race, creed and culture. It is then that your true self can shine forth.

We cannot say it enough, to truly express the Fifth-World consciousness of your evolving planet, all the people of Earth must learn to be brothers and sisters, regardless of your country of birth, your language, your religion or your physical characteristics. Until this state of Oneness is reached, the inhabitants of your world will not attain their full potential or true spiritual maturity

On Free-Will Choice

Choosing to tap into positivity rather than negativity is an active and sometimes difficult choice to make, and something that needs to be worked on with every ounce of your being. Many people do not understand this. They approach life in a passive way, and eventually become overwhelmed by the waves of negativity that are always present on a Level-One planet such as Earth, ready to wash over and drown you if you allow it. For many this negativity becomes a natural way of life, affecting those around them, attracting other like-minded folk to them, drawing even more negative energy into the aura and tainting everything they undertake in their daily life. It becomes a habit, and so much a part of them that they get to a point where they know no other way.

Choosing positivity can be quite a battle at first. It is not passive but rather an *active application of will power*. It is a matter of taking one step at a time, focusing on the present moment and experiencing and acknowledging gratitude for what you do have rather than dwelling upon what you lack. It is also a matter of focusing on others rather than always on self, and simplifying your life so you are not overwhelmed by the complexity and confusion that takes up so much valuable time and energy, stopping you from putting time aside for spiritual nourishment.

Energy is drawn to where mind/awareness directs it, so if the mind dwells on negativity — fear, failure, guilt and worry — that is precisely where all your energy goes. Negative thought patterns are passive and draining, so your energy is siphoned off into nothingness, and nothing will be achieved. On the other hand, when energy is directed into positive and loving thought patterns, it is activated and strengthened, enabling much to be accomplished.

Negativity is complacent and submissive, like a jellyfish being washed to and fro by the tide, with no self-motivated will to move. It is static and non-productive. Positivity must be actively chosen. You are living on a free-will planet, so you must choose to open the door and invite Source to enter, which is an individual free-will choice for each and every one of you to make.

In the Grey cultures there are Councils of Elders to provide advice and guidance, but free-will choices can be made, if these choices don't harm others. Greys are capable of individual thoughts, personalities and ideas, and can choose whether to tune in to the Group Consciousness or not. However, the fact that they communicate telepathically keeps them very honest, as everyone can pick up on others' thoughts, so there can be no subterfuge or deceitfulness. At the same time, it's a point of etiquette to not do this unless invited.

MORE ON FREE WILL

For clarity, let's go into some depth about free will and God's will. Free will is similar to the situation illustrated in cartoons of a person with an angel sitting on one shoulder and a devil sitting on the other, and they're being given a choice to make: Will I or won't I?

Should I or shouldn't I? Can I get away with it or can't I? And this choice can be anything from resisting or giving in to the temptation to steal a candy bar from the corner store right through to embezzling large sums of money or robbing a bank. It all comes down to choosing to act solely for self or for the good of all, and this concept applies across the entire universe.

When you operate through God's will, the temptation to make the selfish choice just doesn't arise. You automatically do the right thing because you are so much an intrinsic part of God/Oneness that you simply cannot act any other way. That is how Zetas and Greys act and think as a collective consciousness, which is quite different from "hive mentality."

Evolution up to these higher levels of the Human Ladder is a return to the Oneness of Source (God). It could be likened to a river system with many tributaries converging as they enter the main river, and that river eventually merges into and becomes one with the ocean, which is symbolic of the God Source. The tributaries represent the thousands of different human life forms that exist on the lower levels of the Human Ladder.

Through the process of evolution, all of these different planetary cultures eventually merge back into Oneness, with each group contributing to the whole. Every planetary culture has its own unique gifts of understanding and knowledge to offer, adding strength and depth to the whole system. The river of consciousness continues to flow through to Level Ten of the Ladder, merging more and more back into the Oneness of the Ocean of Source (God). It is from these broader and more unified levels of the Human Ladder that many of the Creator Beings, also known on Earth as Elohim — which translates as Gods — descend to the lower levels to carry out the creative work of Source. The Greys and Zetas are included in this group, as are many other off-planet and interdimensional beings.

How Will We See Them?

In some cases these beings have biological bodies, but there are also those who have moved beyond the need for gross physical form. However, if their work involves physical space travel and interaction with beings on physical planets they are able to make use of artificial or at least semi-artificial bodies, or what are called "soul containers" (*i.e.*, containers for the soul consciousness). These containers are often based on the Zeta-type form, which is perfectly adapted for interplanetary travel, and in such cases, it's only the body that is artificial. The consciousness that "drives" the body is exactly the same universal Source Energy/Consciousness that drives a biological Earth-human body, which in reality is also nothing more than a "soul container," or vehicle for consciousness. To put this in simpler terms, we can say that some soul containers are biological, and some are not.

The key to universal exploration for many higher planetary groups is the ability to step consciously out of their biological bodies. There are many different atmospheric, climatic and pressure variations on planets, and the biological body you use on your home planet may not necessarily be suitable for another planet, which is no problem if you intend to remain on board your atmospherically, pressure and temperature-controlled mother ship or other craft, but if you need to stay on a foreign planet for any length of time to carry out work, then alternative arrangements have to be made.

You have three choices — placing your disembodied soul essence (your Light Body) into a biological body that is adapted naturally to the foreign planet, which can be done by selecting birth into such a body (as I did in the present life), gaining permission to come as a "walk-in," or making use of an artificial Grey-type body or another one suitable to where you are going. In older planetary cultures these artificial soul containers have been developed to a much higher level of efficiency than are the space suits worn by Earth astronauts. Once you place your consciousness/essence into the artificial one, your energy system permeates it just as it permeates a biological body. In other words, energy-wise you "become" the artificial body, which responds to nerve and brain impulses just as easily and efficiently as does a biological body.

This concept was illustrated brilliantly in the movie "Avatar," in which the lead character goes through a process that removes his consciousness from his crippled Earth-human body and places it into an ET body. The artificial grey bodies are specially designed to withstand interstellar travel. Being artificial, they don't need atmosphere to breathe or food to eat, so they can operate in any planetary conditions.

These artificial bodies/containers were originally developed long ago when the Ant People on Earth were attacked by the Reptoids, who carried out an "ethnic cleansing" on them, which virtually destroyed their ability to reproduce. These artificial bodies provided vehicles for souls who needed to experience life as part of this Earth culture, so they could continue with their planetary ecological work that the Star Elders had assigned to them. These artificial body-forms have since been refined to enable *any* higher soul consciousness to use them by melding their etheric energy systems into the containers, thus providing them with consciousness.

The Zetas operate their ships in the same way, actually linking their energy systems into that of the ship and in so doing become one with the ship. They're lucky in that their bodies, being insectoid and in some cases with a chitinous exoskeleton, are much better adapted to interstellar travel than are mammal-based Earth-human bodies. Grey-type containers are used by many off-planet beings.

WHY DO THE GREYS APPEAR THIS WAY ON EARTH? THEY SHOULD KNOW THAT THEIR APPEARANCE IS GOING TO FRIGHTEN EARTH HUMANS, SO WHY DON'T THEY CHANGE THEIR APPEARANCE AS THEY ARE ABLE TO DO, USING SCREEN IMAGES?

There are several reasons. First, they often do use screen images to lessen the impact of their appearance. Second, why should they change? The bodies that they use are perfectly adapted for the work they're carrying out. That is the way they are, and it's time that humans of Earth stopped judging others by outer appearance and began opening up to the fact that there are others "out there" who do look different. Third, learning to accept others who look different is a major learning curve for Earth humans, and it is part of the maturing process needed for Earth humans to eventually become citizens of the cosmos. Our publisher refers to this inevitable lesson for humanity as overcoming "cosmic racism." Even here on Earth some people fear or dislike other people who look a bit different, but often, once they get to know them, the sense of unfamiliarity dissolves and they become the best of friends.

What about Crop Circle Messages?

We are told to seek three levels of meaning in any message (See "The Three Levels of the Adam and Eve Parable" on page 123), and genuine crop circles are messages from a higher consciousness so we suggest that you look as deeply at them as possible.

For instance, an impressive crop circle appeared near the Torino, Italy, airport June 23, 2015. Around its perimeter was the post-Augustan Latin message *"timeo ET ferentes!"* encrypted in ASCII computer code. Various interpretations of its meaning have been posited. One is, "I fear ET bearing gifts," as in "Beware of Greeks bearing gifts."

Two words in the code are also used in the section of Virgil's Aeneid that tells of the Trojan horse. Consequently some claim that the formation warns us to guard against a false-flag, "Trojan Horse"-style invasion *by evil ETs* who would at first appear friendly, only to destroy us all when our guard is down. From this perspective, this crop circle may have been created by the Controllers to spread further fear and disinformation about "negative ETs visiting Earth," as they often do. They do have advanced technology that can create impressive crop formations.

But the warning may not be about the invasion of evil aliens. There is a second level of meaning. The formation may be a warning about the Earth-based Controllers and the humans they control. Dr. Steven Greer has been increasingly vocal of late warning people of an impending false-flag operation that would try to fool people into fearing all ETs, which would then justify further vast expenditures on space weapons and the like.

The second-level interpretation observes that "et" in Latin can mean "even while", and there was no Latin word or acronym for extraterrestrial. In 2002 the Crabwood formation included an ASCII-encoded disk that was decoded as

"Beware the bearers of FALSE gifts & their BROKEN PROMISES. Much PAIN but still time. BELIEVE. There is GOOD out there. We OPpose DECEPTION."

Capitalization in crop formations often violates standard rules of English grammar, so the use of capitals does not per se require that "ET" refer to extraterrestrials at all. Further, the coded sentence is not imperative, but first-person singular, so it does not warn anyone to fear anything. The second interpretation then becomes, "I (the circle maker) fear, even while bringing assistance," or, more broadly, "I cautiously help you," which is consistent with the Prime Directive interpretation of ETs not forcing any new truths on humans.

CROP CIRCLE, TORINO, ITALY, 6/23/15. A WARNING OF A TROJAN-HORSE FALSE-FLAG PLAN?

A third interpretation has also surfaced: Hundreds of crop circles have had their subtle energies extracted a la Masaru Emoto. The essences created from those energies have been decoded through the Akashic Records, details at NotMadeByHands.com. When the Torino formation was decoded, Akashic Record reader Aingeal Rose O'Grady found "a lot of golden light in it from the beings who created it.... It has a light and joyful energy, with a whirling, spinning motion to it." When she asked who made it, she was told it was created by Arcturians.

HISTORY OF THE GREYS

Like Earth humans and most other planetary cultures, the Greys too have had ups and downs on their evolutionary journey, which goes back billions of years. To quote Kamooh: "No species have been exclusively benevolent or malevolent over the course of aeons, having all been involved in experiencing different levels of consciousness, with their specific karmic burdens included."

Contrary to claims by some channelers and researchers, the Greys are not a mutated Earth-human species, and neither were they created by the Anunnaki as a slave race. Their culture did originally evolve here on Earth, which has given birth to 6 planetary cultures. As with mammalian humans, insectoid humans are now widely dispersed throughout the cosmos.

The Ant People spoken about by Kamooh were ancestors of the Greys. As he explains, they were the first insectoid species to appear on Earth when the original land mass emerged from the oceans that covered the whole planet. These Ant People were then infused with DNA from higher star races, and evolved over eons to reach a high level of evolution and technology, including space travel. Their job on Earth involved interacting with ecosystems to create more diversified environments and life forms, and cultivating organic life, as Zan Tu Kai and I did on other planets in our past Grey life.

To quote Elder Kamooh: "The Ant-People were created by the Star Elders to assist them in their work of caretaking the garden (Earth). Their spiritual mission, like their Star Elders, was to continue encouraging inter-species Soul relations. Introducing new species or hybrids to environments, they accelerated biological symbiosis evolution. With the help of the Elementals, their Elder brothers, they participated in developing biodiversity."

Like ants and bees, the Ant People operated as a hive consciousness, depending upon a female "queen" to reproduce by laying eggs. They reproduced rapidly, becoming a major threat to the Reptoids, who considered themselves the dominant species and owners of Planet Earth even back then. They became caught up in major wars with the Reptoids, who tried to wipe them out by destroying the female "queens" and driving their colonies deeper and deeper underground, then off the planet. Many who survived the attacks were enslaved, genetically interfered with, and further hybridized with reptilian genetics. They were controlled by implants placed into their bodies by the Reptoids, as is happening now with many Earth humans. A form of "ethnic cleansing" was carried out to stop them from reproducing normally.

The Ant People/Greys eventually left Earth to take up residence inside the present moon, which they constructed for this purpose and where they still have bases. They also went to other star systems such as Arcturus and Zeta Reticuli, where another major war occurred when they were again attacked by the Reptoids and their "Blond" allies.

Given their past history, the Greys understand intimately how steep and slippery the path of spiritual evolution can be. Like Earth-plane humanity, they too once reproduced to the detriment of their society. They also suffered a loss of their emotional well-being through technology, as the Reptoids played "divide and conquer" games with them, as is now happening with Earth's people. The Greys have fallen by the wayside 3 times, and have gotten back up again each time.

To quote from remote viewer Courtney Brown's book, Cosmic Explorers, in which he discusses the Greys' spirituality: "Interestingly, however, though they lack for nothing physically, it is spiritual growth that they seek most strongly. Never have I encountered a species in which the quest to evolve towards God is felt so deeply. In my view, this species looks toward spirituality like a starving man looks at an apple."

It truly is almost an obsession with us! And no wonder the Greys are saddened by what is happening on Earth! Yes, there has been the need for an exchange between our two cultures, with some of our people requiring Earth-human genetics to assist them to regain their human-ness, in exchange for repair and activation of Earth peoples' DNA to assist in the step up to 5th World Consciousness. Speaking as a Grey, all I can say is - we do understand what Earth-plane humanity is up against. We know where the pitfalls are, having "been there, done that" ourselves. It breaks our hearts to see this happening yet again to our beautiful Mother Earth and Her people. We are not an "evil alien force come to invade your world." We are from here originally, and once we healed our culture, we became citizens of the Universe, as you can too!

PEOPLE OFTEN WANT TO KNOW PHYSICAL FACTS ABOUT THE ZETAS AND GREYS SUCH AS—

Are their ships actual physical vehicles or is this just an illusion?

The ships can be actual physical "nuts-and-bolts" objects, or they can vibrate on a higher frequency beyond physical reality.

Advanced ET technology and expanded conscious awareness enable the vibrational frequency of both animate and inanimate objects to be altered at will, depending upon what is required at the time. In the Greater Reality, "reality" is subjective rather than objective, and therefore it is not as rigid and fixed as *it appears to be on Earth.*

This is the point that seriously blocks some UFO/ET researchers on Earth who refuse to look past the "nuts and bolts" aspect. They are missing the vital component of consciousness-enhanced and multidimensional technology.

Do they eat food like we do, or not?

Food requirements depend upon which level of the Human Ladder you're consciously operating on. In Zeta/Grey reality, quite a range of levels and energy frequencies are spanned, so some Greys need more physical food/nourishment than do others. For example, an Essassani Zeta would require more physical nourishment than would a tall Grey, because they operate on a physical level, whereas the tall Greys are of a higher, less physically oriented frequency.

Those at the top of the Ladder don't require physical nourishment at all, as they are pure energy beings, and beyond the grossness of physicality. Greys who do eat are mainly vegetarian, and their food is prepared as a pudding-like substance to be eaten, or as a paste that is rubbed into the skin.

Suzy Hansen has been involved in UFO research and sighting investigations for 40 years. In 2000, she founded the UFO Focus New

Zealand Research Network. She describes being taken into a food preparation area on a ship, where several short Greys were preparing wafer-like food for the ship's crew. The Greys cannot chew food as can Earth humans, so the wafer is either placed directly onto the skin or into the mouth to dissolve (like the communion wafers used in church services down here).

The Greys are totally egalitarian, so all crew members on a ship take turns at "kitchen duties." Waste matter is cleared through the skin or breath. On the huge Mother ships, all levels of manifestation need to be catered for, because many different planetary beings may be on board at any given time, so therefore physical requirements such as food preparation, waste disposal, bathrooms, living and sleeping quarters, *etc.*, must be provided.

Do they live in houses like regular humans, and if so, what are these dwellings like?

In the Zeta Reticuli system, some people live in apartment-type complexes under protective domes, known as vadins, while others live in more natural environments. Helene was once taken on an out-of-body journey by one of the Grey Elders, and shown a complex made up of rounded, brownish-colored, dome-shaped, cavelike dwelling. Reading Kamooh's description of the dwellings of the Ant People, the ancestors of the Greys who lived on Earth long, long ago, reminded me of this observation by Helene. Kamooh explains that they made their homes in such dwellings or else in deep underground caverns to keep themselves safe from the reptoids. I'm also aware, from more recent Grey lives, of time spent living on a ship like a travelling city, following a sort of "cosmic Gypsy" lifestyle of travel and exploration.

Suzy Hansen describes the Greys' living quarters on a ship exactly as I remember them, quite independently validating each other's recall. I'm in contact with several blended-soul Greys down here in Earth-human form, and all of us agree on this point, being strongly drawn to the softly flowing, pure-white architectural style that is a feature of Greek Islands such as Santorini and Mykonos.

Understanding the Grey Guardians

This style clearly reminds us of our living quarters on board the ships, which consist of small, white, almost hivelike rooms that open onto arched passageways. Everything is rounded and "molded," without sharp corners or straight lines. Centuries of whitewashing gives Mediterranean architecture a similar appearance.

Each room is fitted with what Suzy refers to as a "shelf bed," which resembles a thin plastic shelf jutting out from the wall, but incorporated seamlessly as part of the wall. These are extremely strong and comfortable, and are linked into the consciousness of the ship, so they mold themselves into the exact shape of your body when you lie on them, then they flatten out automatically when you get up, so there's no need to "make the bed." Bed clothes are not needed, as the ship is temperature controlled.

These beds have a balancing, healing and harmonizing effect on the body and energy system, so if you've finished a hard shift and are in desperate need of some R & R, these beds are perfect. When you go to "sleep" on them, you actually merge back into the soul consciousness of the group and the ship, which basically "defrags" your consciousness and rebalances your energy system.

As Suzy also describes in detail, the only other item of furnishing in each room is a "virtual reality" chair, which serves their entertainment needs. One can sit in this chair, which has a screen that comes down in front of your face, with controls on the armrests. This chair provides a total 3-D reality experience (and beyond) of whatever you choose, so you can watch a movie, travel to another world, place yourself in any part of the ship, or whatever. As it is a virtual-reality experience, you don't just see what's going on — you can smell, touch, taste and hear exactly as if you were right there.

Do They Have Families?

The Reticulan and other Grey cultures are very family-oriented, with an official ceremony that is performed to join a couple together in the equivalent of marriage. It's not a physical joining but rather a linking of the partners' auras, so it's very much about energy melding. Only one or two children are allowed per couple, and because

Zetas can no longer reproduce sexually like Earth humans, reproduction is carried out by both partners providing genetic material which is spliced and grown *in vitro* to produce a child. In this way cloning can be avoided, as both partners contribute to the child's gene pool. In the case of Essassani hybrids, Earth-human and Zeta genetics are combined, and the Earth-human parent, generally the mother, is brought up onto the ship to bond with the child.

Most planets have much stricter laws than Planet Earth. Planet Earth is very much a "school planet" where newly evolved humans come to learn how to handle free will, so therefore a much wider range of both positive and negative choices are available, so down here you either "sink or swim." Everyone learns to "swim" eventually. On planets in Reticuli and most other star systems, reproduction of the human inhabitants to the detriment of other creatures co-inhabiting the planet is strictly forbidden. All creatures are recognized as expressions of Source/God Energy, and therefore must be respected. Other higher-ET societies are horrified at how Earth humans are permitted and, in fact, encouraged to breed in such large numbers as to cause major damage to the whole ecosystem of the planet. All creatures are an aspect of God/Source and therefore should be respected as such.

Elder Kamooh: "The act of reproducing without this (Soul) consciousness, as you are pressured to do by the lower lords, can give any odd result and allow any soul to incarnate, some who can be related or not (in) helping your individual and collective evolution. This explains the reason why of all species, *only yours* seems to overpopulate and take over all others. *Only spiritual dharmic conception can solve your problem*, meaning that when you are ready to conceive and give birth to a soul, you do it for the Greater Soul." (my italics)

Kamooh: "The long story of the spiritual evolution on our Earth and in the Omniverse teaches that struggles and conflicts are part of the learning and growing process, and the contrary energies also have a purpose. We (the Sasquatch) have grown up in our understanding through experience, including the wars with the lower powers. We learned from them what to avoid, how to be brave and faithful, confirming our own sacred mission. We know that the opposition of forces is part of evolution, making us grow stronger, clearer and wiser. Our Soul will always be tested, as we stand strong as spiritual warriors protecting the Cosmic order, or if we lose track of our ancient star seeds memory, disconnect from our spiritual evolution purpose, get entrapped in materialistic frames, enslaved by the controllers destroying our world, life and souls."

Remember Kamooh's wise words: "Every soul and entity is always offered the possibility to change their way, heal and evolve, as long as their consciousness doesn't regress into eternal lethargy, which is spiritual death."

28

Final Thoughts

"Be the change you wish to see in the world."
— *Mahatma Gandhi*

As recent ETs, many of us volunteer to partake of a lifespan as a human on Planet Earth. We do this to specifically enable us to experience fully what it's like to actually be an Earth human, confined to three-D reality and operating through a Level-One, 10 percent active consciousness. In this way we gain a deeper understanding of the difficulties involved. Fear is the main issue on Earth, so it's important for us to experience it as fully as Earth humans experience it, but, by doing so, we open ourselves to the possibility of negative interference.

Some of us have temporarily forgotten our mission on this planet and are not consciously aware of who and what we are, but nevertheless, we can feel very down-hearted, lost and homesick at times. Because of the limits of our Earth-human awareness, we often don't understand what it is that's affecting us so deeply. The important thing to remember is that living a life on Planet Earth in human form is a very short mission compared to the bigger picture of eternity. After all, the span of a human life is only a tiny droplet in the vast ocean of eternity, and it is an incredible honor to be here at this time, helping humanity towards Cosmic Citizenship. Even if we aren't consciously aware of who and what we are, our presence alone is helping Planet Earth and her people on an energy level to become more self-empowered.

You are reading this book, so it's very likely you also have off-planet connections and may be one of the many volunteers who have chosen a life on Planet Earth as part of our "rescue mission."

Final Thoughts

If you can just take a little time to dig deep enough to ask yourself some questions, some memories may come to the surface.
- Do you feel a connection with any particular off-planet group?
- Have you had any visions or dreams of being on board an ET ship, or having seen a ship in the sky?
- Perhaps you've received healing or teaching from your off-planet family, or like me, have healed others in your ET form.
- Could the fact that you're here be simply a part of your mission as one of the off-planet volunteers?

If you are feeling homesick and lonely for "someplace else," once these memories and realizations come to the conscious level you'll feel better, because you'll suddenly find much more purpose and reason behind your existence on this planet. You'll also come to know that you're not alone. There are many of us on this mission, Zetas, Pleiadians, Lyrans, Sirians and many, many others, all working together in perfect love, harmony and oneness, supporting and caring for each other.

I'd like also to add a word of reassurance here. Some of you will feel familiarity and kinship with one or more of the many Reptilian cultures from off-planet. Don't dismiss this! These folks are some of the best healers in the universe, and they also act in a protecting role for other star people. The only negative ones are those based right here, along with the fake Greys, who are *not* ETs. The off-planet "Lizzies," as our Pleiadian friends jokingly refer to them, are kind and good people, and very much a part of our team.

Some years ago I exchanged emails with a woman named Jujuolui Kuita (Juju for short), who is one of the hybrids featured in Miguel Mendonça's two books, *Meeting the Hybrids*, and *Being With the Beings*. Juju has a strong connection with the Fajan people, a Reptilian race based on a planet called Faqui. As told in *Being With the Beings*, Juju is a former soldier and police officer who now devotes her life to animal welfare. After being involved in rescue work after Hurricane Katrina hit the southern states of America, she was led to take courses in disaster response so that she could devote her life to being more "of service." She now works as an Emergency Animal

Medical Technician, using her skills as a Reiki practitioner to help the animals she rescues. Juju describes her Reptilian culture as evolved, peaceful beings, with great love and compassion. She is a perfect example of these virtues.

Shades of Grey

I am well aware of the massive amount of negative information put out on the "Greys," and also that some people have reported unpleasant experiences in connection with them. We must clarify the fact that two *very different* groups of "Greys" are involved. One group is comprised of highly evolved insectoid humans from off-planet, and the other group consists of cloned insectoid/reptoid hybrids resident here on Earth and on the moon. There are similarities in appearance between them, which has been done on purpose to cause confusion.

In her books, Sanni Ceto speaks of interplanetary wars that occurred eons ago involving a Zeta Reticulan planet that was attacked by Reptilians and their "Blond" allies, a race of Reptoid/Pleiadian hybrids. The invaders abducted the planet's inhabitants and forced them to crossbreed with reptoids, who fitted them with implants to control them and then used them as slaves. A similar event occurred on Earth involving the Reptoids and Ant People.

The Ant People were on Earth billions of years ago. They are very distant relatives of the off-planet Greys in the same way that Cro-Magnons and Neanderthals are distant ancestors of modern-day humans. Having evolved from ants, the Ant People operated as a hive consciousness, and, as an early humanoid species, they had their wars along with a rather polarized outlook, as do many of today's Earth people. As with present-day humans, some were more highly evolved than others. The Repterrans were also present on Earth as "Controllers," just as they are now, so even back then Earth was a war zone.

When some of the Ant People were captured by the Repterrans, they, like the captured Zetas, were crossbred with reptoids, fitted with implants, cloned and enslaved, which stopped them from evolving as have the genuine Greys. This is still being done by the Controllers. These cloned insectoid/reptoid hybrids are the PLFs (Programmed Life Forms) that Steven Greer refers to. They are not Greys, but they do slightly resemble Greys. More positive Ant People are still present, mainly based on the moon. The more evolved off-planet Greys and other ETs are trying to assist these Ant People to evolve as they are also doing with Earth humans who wish to move on spiritually, hence the DNA work being done with both.

Off-planet Greys began returning to Earth after the detonation of the nuclear weapons that ended WWII. They did use the moon as a base, as some of the Ant People are settled there. Some Ant People were present on off-planet ships that had crashed. The off-planet people coming here were deeply concerned, as were the Earth-plane Ant People, that humanity was, and is, taking Earth down a road to destruction, especially once nuclear energy started being developed. The ships recovered from crashes were not nuclear-powered. It took many years of intensive back-engineering to even approach an understanding of off-planet technology, which is still not fully understood, as consciousness-enhanced technology is involved. The Nazi "bell" was a crude replica of an ET craft.

Some Common Myths that need Busting

Myth — The tall Greys are Reptoids

Reality: The tall Greys are not Reptoids, rather they are highly evolved Elders of the genuine Greys. They are Insectoids, either Zeta Kebbans or Mantids. I and many others have had ongoing contact and guidance involving the tall Greys since early childhood, and that contact has been totally loving and positive. Extensive and very comprehensive surveys carried out by the Dr. Edgar Mitchell Foundation for Research into Extraterrestrial Encounters (FREE) provide well-corroborated evidence that the *vast majority* of ET contact is positive.

Myth — The Greys are carrying out ecocide on Earth.

Some people believe and spread rumors that the Greys are planning to destroy humanity so that they can take over the planet. Part of their supposed plan is to encourage the carbon levels to rise to return Earth to the hot climate that existed eons ago.

Reality: The FREE survey concluded that the majority of people contacted by Greys and other ETs have become much more environmentally minded and feel a higher level of care and empathy with Mother Earth and Nature. Why would the Greys and other ETs urge people to become more environmental if they wanted to destroy it?

As for the false claim made by some that the Greys are encouraging higher carbon levels and global warming? People taken up onto ships, particularly those of the Greys, find the temperature uncomfortably cold. The Greys do not like heat, whereas the Reptoids do. It's the Repterran Controllers who are behind the denial of global warming.

> *Myth — The Men in Black are tall Greys disguised to appear human.*

Reality: The Men in Black are reptoid hybrids created on Earth by the Controllers. They are created as they are to cause confusion. The genuine tall Greys are often seen dressed in black clothing, because their vibrational frequency is so high that it doesn't reflect light waves. This is similar to so-called dark or black matter, which vibrates beyond the speed of light. In their own environment up on the ships, the tall Greys dress in white robes.

Since the publication of my first two books, I've received many emails from all over the world from Greys, Mantids and Reptoids who are based here as part of our mission. All of them, with only one exception, have been kind, caring beings who simply are dedicated to doing what is best for Earth and humanity. Many of them, like Juju, have taken on caring roles as healers, teachers, rescue workers, paramedics *etc.*, to enable them to be "of service" in the most effective way they can. Like me, they have their off-planet family as well as their Earth-plane family, and all any of us want is peace and a safe and healthy environment, on Earth and throughout the universe, to enable us to live out our lives to the best of our abilities.

The Mantids are Elders to the Greys. When Oris, Maris and Serapis, the tall Grey Elders with whom Helene and I work, first introduced themselves to us, we wondered why all their names ended with the "IS" suffix. I thought that it was just a Grey-teacher "thing," and thought no more about it. However, we eventually came to understand that it is their way of identifying themselves as Mant*IS* beings; the tallest and most highly evolved of the Grey Elders. As strange as they may look, they are incredibly kind and compassionate, and have a great sense of humor. This applies with the Reptilians as well, who are extremely protective, hence the roles they often take on down here as emergency and rescue workers. They are often big, strong people who thrive on such active and often dangerous work. The important thing to remember is, "beautiful" does not necessarily equate with "good," and "ugly" does not necessarily equate with "bad." *Discernment of the energy field surrounding any being, on or off-planet is the key.*

Importance of Discernment

There is so much information and disinformation being put out, in seemingly ET and spiritual contact with humans, conspiracy theories, exopolitics, *etc*. How can we sort out truth from fiction? How can we tell if information we're being given is positive or negative? I once put this question to Maris, my Grey Teacher, and his answer was: "Try to sense the energy behind the words, whether they are written or spoken, and ask yourself — is this message coming from love or fear?" It's so important to practice discernment in your dealings with others, whether they're Earth human, ET, angelic or whatever. It has nothing whatsoever to do with how the being looks or how beautiful their words sound. It's all about the energy behind the words and what you sense in their aura.

"People are just now beginning to recognize the significance of the new alliance that is building between Earth humans, who need help, the Guardians from Zeta, who want to help, the Earth ETs, who are standing by, and the never-controlled by-Controllers Sasquatch People.

"A new world is upon us, and it is very hopeful. Take heart and pass the word carefully."

—Kamooh, Sasquatch Elder

Glossary

3-D awareness: perceiving what we see as real rather than as a holographic projection from a higher-dimensional reality. Humans usually see in only three dimensions plus time. p. 17

ACERN: Australian Close Encounter Resource Network p. i

Alcheringa Adaptation of the name Alchquarina, the Commander-in-Chief of an ET ship occupied by Pleiadians and Lyrans who became stranded in Australia 900,000 years ago after their ship crashed. p. 63

Angels: very highly evolved soul essences operating at the top of the Human Ladder. They have moved way beyond the need, or even desire, for physicality, and are more than 90% consciously aware. p. 29

Anunnakis: Repterrans who founded the Sumerian and Babylonian cultures p. 74

Ascension: raising our vibrational frequency to higher and more refined frequency bands. p. 147

Astral plane: a band of non-physical dimensional frequencies that was originally created as a "holding ground" for Earth-human consciousnesses while humans are out of physical form between lives, sometimes called the bardo. It is composed entirely of the collective consciousness of Earth-plane humanity, along with myriad thought forms, both positive and negative, that have been created through all the many belief systems of Earth. p. 22

Baime: the first Creation Spirit of Aboriginal lore p. 114

Blonds or Nordics: an Inner Earth group who inspired Hitler's Aryan Ideal and drove his obsession to eradicate certain racial types on Planet Earth and replace them with this race of super beings. p. 88

Glossary

Brain parasite: introduced by the Te-raks, which fed the negative-emotion centers of their brains, causing them to become hostile and controlling, thriving on war and aggressive physicality. p. 68

Chakras: vortices of subtle energy in the energy grid system of the body. There are seven major chakras and many minor ones. p. 132

Conscious dying: carries consciousness out of soul containers to prepare for the next life. p. 144

Controllers: an Earth-human group descended from a technically advanced off-planet group that settled on Earth millions of years ago, long before Earth humans were here. They refuse to relinquish their ownership and complete control of Earth and humanity. See also Repterrans. p. 66

Created in the image of God: in reality humans were created in the image of the Repterrans themselves. p. 52

Crop circles: freely bestowed symbolic gifts to humanity and Mother Earth from the Assistant Creators. They are True or Universal Symbols that both invoke and evoke a high-frequency energy that links directly into human consciousness to assist in the ascension of both Earth and its people. p. 33

Daughters of Men: the female offspring of the new human species. p. 124

De-raks: biped human/reptilian beings developed from dinosaur stock by the Draconians. p. 67

Doing: the creation, development and evolution of all life forms throughout this universe. p. 16

Draconians: a reptoid human species native to the Dorado and Pictor sectors, with some colonies scattered in the Orion Nebular sector. Original settlers of Earth, Assistant Creators who reseeded the planet with dinosaurs following an early cataclysm on Earth that had destroyed much of its biodiversity. p. 66

Dreamtime: Aboriginal legend firmly based on the story of Ancestor/Creator Beings who came to Earth from off-planet. p. 48

Drisei: a more evolved reptoid culture. p. 64

Emotion: A form of energy vibrating at a certain frequency. p. 176

Energy: is like water that seeks a balance p. 23

Escherichia coli: E. coli bacteria, an ancient and primitive organism with no nucleus. Perhaps the brain parasite that has infected the Earth for millennia. p. 68

Essassanis: a mammalian/insectoid hybrid group living in the Zeta system of mixed Zeta and Earth-human genetics. p. 65

Evolution: a gradual speeding up, expanding and refining process through which energy passes, or "ascends." p. 136

Fajan people: a loving and compassionate Reptilian race of evolved, peaceful beings based on a planet called Faqui. p. 208

Faqui: home planet of the Fajan people, a loving and compassionate Reptilian race of evolved, peaceful beings. p. 208

Fifth World: According to ancient Hopi legend, our current (Fourth World) of competition and war gives way to a peaceful and balanced Fifth World sometime around now. See bit.ly/2itnLBX p. 35

FREE: Foundation for Research into Extraterrestrial Encounters p. i

Fruit of the Tree of Knowledge: reasoning power, or "knowing," of which "Adam and Eve" partook prematurely. This reasoning ability is part of the evolutionary path from the purely instinctual drive of the animal kingdom to the reasoning power of the human, but it requires free will. p. 49

God of the Old Testament: was actually Repterrans trying to keep the New Humans under control. These Repterrans are prone to the very human weaknesses of jealousy, anger and the need for praise and flattery. p. 51

God: the source of universal energy, is of the highest vibrational frequency, manifesting as Love and Light p. 137

Heart chakra: bridge and balance point between upper and lower chakras p. 179

Human Ladder: a term used by the Greys for ten ascending human evolutionary levels in the universe. p. 63

Hybrids: ETs who have some Earth-human genetic material in their makeup p. 188

Indigo, Rainbow and Crystal Children: more recent generations of genetically upgraded children who have been born on Earth over the past few decades p. 185

Glossary

Inner Earth: a multidimensional realm that spans both physical and astral planes and which has been inhabited for eons by Repterrans, some of whom are Controllers. p. 86

Interdimensionals: universal caretakers and custodians whose job is to oversee the development of life on many planets. p. 53

Isis, Egyptian goddess p. 122

Kariong: site of ancient hieroglyphics from survivors of the Rexegina. north of Sydney, New South Wales, Australia. p. 46

Kebbans: Zeta Reticulan Greys who, like the Mantis people, are also a highly evolved, gentle and peace-loving society. They are highly attuned to Universal Oneness/Source and are also Grey Guardians. p. 65

Love: another name for God p. 163

Lyrans: a race closely related to the Pleiadians and Earth humans. Cousins to Earth humans. p. 62

Made in the image of God: Source Energy permeates every level of creation from the largest galaxy down to the smallest part of every atom. p. 18

Mantis beings: the tallest and most highly evolved of the Grey Elders, kind and compassionate, with a great sense of humor. p. 212

Merope: a forested planet in the Pleiades, with abundant water and a climate similar to that of Earth but with less severe winters. Beings from Merope are called Meropians. p. 62

More highly evolved ETs: universal caretakers and custodians whose job is to oversee the development of life on many planets. p. 53

Multidimensional: having an existence beyond the normal three dimensions and time that humans experience. p. 137

Multiverse: many universes existing simultaneously p. 137

Near-death experience (NDE): when physical trauma to the body induces an out-of-body experience. p. 148

Nephilim: The giant, half- reptilian/half-mammalian offspring resulting from Fallen Watchers taking the "daughters of men" (New Humans) to themselves as wives. p. 50

New Paradigm: New Science is telling us that the Old Science model is not correct; perceived reality is not true. We are not observers; we are participants. Our observations, perceptions, and intentions exert a slight but measurable influence on the world within us, as well as the world around us. p. 35

Oneness: The concept that we are all One, Earth humans, Zetas, Pleiadians, Sirians, etc., and all life forms in the universe. p. 16

Orb: a pure soul essence manifesting free of gross physical form. p. 28

Out-of-body experiences (OOBE): when one's awareness temporarily disconnects from the physical body. p. 148

Past-life memory: when one has a clear and deeply emotional impression of having lived in another time and place. p. 148

Pharaohs Akhenaton and Tutankhamen, off-planet people who came down here in hybrid containers chosen specifically at that time to enable them to remain for the span of a lifetime in human form. p. 56

Physicality: operating in 3-D perceptual reality. Only a small part of the universe vibrates at gross matter levels like the Earth — most of the universe is beyond physicality. p. 29

Planetary energy grid: a matrix of subtle-energy lines that influences the entire planet p. 132

Pure-blood Zetas: one of several cultures referred to by Earth humans as Greys. p. 65

Puy-yats: biped human/reptilian beings developed from dinosaur stock by the Draconians. p. 67

Rainbow Serpent: a motif in Aboriginal art that is representative of the DNA and genetic work carried out by Ancestor/Creator Beings whose likeness is recorded in the sacred Wandjina drawings. p. 48

Religion: a belief system based on dogma, with an intermediary priesthood and set rituals.;spirituality a more personalized state of consciousness. It is about trying to the best of your ability to practice unconditional love and do unto others as you would have them do unto you. p. 152

Reprogramming: realigning human energy systems to suit a given purpose. p. 132

Glossary

Repterrans: Reptoid humans developed from the dinosaurs on Earth by the Draconians. Repterrans include the De-raks, Te-raks and Puy-yats. They are highly advanced in the field of genetic engineering and consider themselves to be the dominant human group and owners of Earth. See Cabal and Controllers. p. 67

Reptilians: off-planet beings incarnating in soul containers with reptilian genetics. They are some of the best healers in the universe and act to protect other star people. p. 208

Reptoids: beings derived from reptilian genetic stock p. 67

Rexegena- Pleiadian mothership that was sent to Earth 900,000 years ago with 50,000 people on board consisting of a large crew and their families. p. 44

Safe zone- the vibrational frequency around an inhabited planet that stops less-evolved groups from attacking the planet's inhabitants. p. 63

Satan: the energy of fear within human hearts. p. 129

Sector- a spatial volume that includes many star systems. This is a designation Earth humans do not generally use. p. 66

Serpent in the Garden of Eden: represented the rebel Repterrans who tempted them with the forbidden fruit — the ability to reason along with the gift of free will, neither of which they were ready to handle at that stage in their development, being still too immature on the spiritual level to handle such abilities wisely p. 128

Shelf bed: a thin plastic-like shelf jutting out from the wall of the crew quarters of a ship, but incorporated seamlessly as part of the wall. They are linked into the consciousness of the ship. p. 204

Soul container: a physical body that contains the spiritual essence of a being. The Grey type of soul container, is simply a temporary vehicle used by the soul essence that is perfectly suited for inter-planetary travel on missions which involve interaction on physical third-dimensional planets such as Earth. p. ix

Soul socializing: A Zeta term for having deep and meaningful conversation with others p. 158

Soul: our immortal, eternal spiritual/soul essence that is independent of the biological body p. 148

Source: name for God p. 16

Source: name for God p. 17

Star Children: Earth children who have been energetically upgraded with ET genetics and enhanced DNA. p. 188

Supernatural: aspects of nature that until now have been beyond the limited understanding of science p. 136

Te-raks: biped human/reptilian beings developed from dinosaur stock by the Draconians. p. 67

Ter-hig-gom or Tar-hig-gom: a more evolved reptoid culture p. 64

Terra, or Terai: Earth in the Zeta Reticulan language p. 67

Terraform- to manipulate the environment of a planet over long periods so as to make it habitable. p. 62

The All: a name for God p. 17

Three Sisters: Three organizations that comprise the Cabal, i.e. the shadow U.S. government run by the Repterran Controllers: the Council on Foreign Relations, the Bilderbergers and the Trilateral Commission p. 90

UFOs: (Unidentified Flying Objects): flying objects not yet understood by human authorities p. 153

Unconditional love: expressed through compassion, patience and peace. p. 163

Universe: energy vibrating across infinitely many frequency bands of the electromagnetic scale. p. 137

Upstairs: an informal reference to all the seen and unseen allies that guide humans p. 83

Vadins: apartment-type complexes under protective domes in the Zeta Reticuli system. In the Romany (Gypsy) language, which is based on Sanskrit, the ancient language of India, the word for their traditional home on wheels is "vardo," which derives from the Reticulan word "vadin." p. 203

Xenophobia: the fear of anything or anyone that seems different p. 182

Yang and yin; positive and negative; male and female, opposite polarities in Chinese medical/spiritual philosophy. p. 124

Zeta Reticulans: a race of intelligent beings who befriend the Earth, often visiting here in their short gray-skinned soul containers with large black eyes and oversized head. Author Judy Carroll's Earth-human body is occupied by a reincarnated soul who previously occupied a Zeta soul container that was ruined in a vehicle crash in the US southwest in the 1940s. p. 74

Resources For Further Research

If you would like to read more on the various interplanetary races, their cultures and their home worlds, I highly recommend Sanni's two books — *Stranded on Earth: The Story of a Roswell Crash Survivor* and *Zeti Child: Lost Upon a One-Star World*. You may also read about Sanni on our website — www.ufogreyinfo.com

For an extensive interview of Jacquelin by renowned researcher and author Barbara Lamb, see her new book, co-authored by Miguel Mendonça, *Meet the Hybrids: The Lives and Missions of ET Ambassadors on Earth*.[1]

Many crop circle images and information can be found at web sites run by the Crop Circle Connector, Brett Parrott, and NotMadeByHands.com, a site that has collected and decoded through the Akashic Records life-force energies from many crop circles and sacred sites. (See "Names and Works" on page 227.)

If you wish to read more on quantum physics, Intelligent Design as opposed to Darwinism, and the ability of the human mind to create its own reality, I suggest this list of books and presentations provided by Dr. Leo Sprinkle under the heading "Old Science and New Science" on p. 11 of his presentation: "ET Experiencers: From Planetary Persons to Cosmic Citizens?"

Dr. J. Bockris, Ph.D., (2007), a physical chemist who has presented *The New Paradigm: A Confrontation between Physics and the Paranormal Phenomena*.

Dr. Amit Goswami, Ph.D, a professor of physics, (1993), has presented *The Self-Aware Universe: How Consciousness Creates the*

---

1. Jacquelin's website: jacquelinsmith.com

RESOURCES FOR FURTHER RESEARCH

Material World, and, (2001), *Physics of the Soul,* which presents a model of reincarnation.

Suzy Hansen's book *The Dual Soul Connection: The Alien Agenda for Human Advancement* presents a complementary view of life as an Earth ET, a human body occupied by a soul not from Earth.

Dr. Michio Kaku, (1994), has presented a brief but lucid history of modern physics, *Hyperspace: A Scientific Odyssey through Parallel Universes, Time Warps, and the Tenth Dimension.* Charles T. Tart, Ph.D., has argued that the evidence for paranormal phenomena is bringing science and spirit together.

Dr. Dean Radin, Ph.D, (1997), a neuroscientist, has analyzed ESP experiments in his 2006 book, *The Conscious Universe,* recently expanded his analysis of "quantum reality" and the implications for "entangled minds."

Peter Russell, (2002), has described his journey as a physicist in his book *From Science to God.*

Dr. Gary E. R. Schwartz and Dr. Linda Russek, Ph.D, (1992), have written *The Living Energy Universe.* Also, Dr. Schwartz and colleagues, (2001), have demonstrated scientific evidence for life beyond human existence in *The After Life Experiments.*

SunBôw, as told by Sasquatch Elder Kamooh, relates the 50-million-year history of the Sasquatch People on Earth in *The Sasquatch Message to Humanity: Conversations with Elder Kamooh. (See below.)* Their DNA contributed to that of *Homo sapiens,* making the Sasquatch like a big brother to the human race. The seven races of the Sasquatch are highly telepathic and interdimensional and have never been controlled by what they call the "Lower Lords."

SunBôw, *The Sasquatch Message to Humanity: Conversations with Elder Kamooh, Book 2: Interdimensional Teachings from Our Elders.*

Claude Swanson, Ph.D, physicist, (2008), has written *The Synchronized Universe: New Science of the Paranormal* and *Life Force, the Scientific Basis: Volume 2 of the Synchronized Universe.*

For anyone still entrenched in fear and feeling victimized by ET contact experiences but wanting to move on, I recommend *The Forgotten Promise* by Sherry Wilde.

For more translations of ancient texts from Aramaic to English, see DaleAllenHoffman.com/

SASQUATCH

Deep and special acknowledgement needs to be given to author SunBôw, Sasquatch Elder Kamooh, and to author Kewaunee Lapseritis, a world authority on Sasquatch phenomena.

When publically presenting controversial material to the public, it is advantageous to have as much validation as possible from independent sources. Such validation recently turned up on the desk of my publishers, in the form of a book titled *The Sasquatch Message to Humanity: Conversations With Elder Kamooh,* written by SunBôw, a French Canadian with Métsis ancestry. SunBôw has spent a lifetime of spiritual study and journeying with the elders and shamans of many Native American Nations. He has also had ongoing contact and received teachings from an ancient Sasquatch Elder by the name of Kamooh. In the words of my publisher, this book fits hand-in-glove with the information I am presenting here in *Extraterrestrial Presence on Earth*. *The Sasquatch Message to Humanity: Book 2: Interdimensional Teachings from our Elders*, was released in 2017.

Both SunBôw and Kewaunee Lapseritis have kindly given me permission for material from *The Sasquatch Message to Humanity* to be reproduced in this book. Kewaunee's websites are sasquatchpeople.com and psychicsasquatch.com, where all 50 chapters of SunBôw's book can be read.

Links provided by SunBôw - Ironwood Log Project, on Isuma Native online TV: isuma.tv/fr/ironwood-log-project

Excerpts from interview on Prophecy Keeperz radio:
  YouTube.com/watch?v=mFWPS-7e-7E

One United Nation for the Great Earth Peace:
  facebook.com/groups/oneunitednation

Figures

Sanni Ceto	p. 10
Jacquelin Smith	p. 11
Helene Kay	p. 12
Valerie Barrow	p. 13
Explanation of the Human Ladder	p. 26
Soul-Essence Orbs in Judy's Back Yard	p. 28
Milk Hill Formation 8.13.2001	p. 30
Milk Hill Formation in the Aboriginal Style	p. 32
Illustration of Schumann Resonance	p. 36
Aboriginal Petroglyph, Elizabeth Station	p. 47
UFO Sand Sculpture at Mt. Kurama	p. 57
Australian Art	p. 109
Mayan artifact of a Grey beneath a craft	p. 110
Natural Rock Formation at Uluru	p. 112
Scribbly Gum Tree	p. 115
Grandmother Tree	p. 116
Entrance	p. 118
First Vortex	p. 119
Second Vortex	p. 119
Glyph of a pregnant star woman.	p. 121
Isis Grey	p. 131
Ghirlandaio's Madonna w/ St. Giovannino	p. 183
Crop Circle, Torino, Italy, 6/23/15	p. 199

Names and Works

A
Adam and Eve v, 128, 129
   emulated the Repterrans 130
Alchquarina 45, 110
   Lyran Commander-in-Chief of the Rexegena. 45
Alexander, Dr. Eben
   Proof of Heaven- A Neurosurgeon's Journey into the Afterlife 68
Arthur C. Clarke
   tinyurl.com/j3o8t7d 51

B
Barrow, Valerie 113
   Alcheringa ii, 63, 110
Bockris, J., Ph.D.
   "The New Paradigm A Confrontation between Physics and the Paranormal Phenomena." (2007) 223
Braden, Gregg 36
   Fractal Time and others 35
   Human by Design, forthcoming 49
Bruno, Giordano
   suppressed by Controllers 59
Buddha
   Guardian who attempted to reverse Reptilian control. 56
Buddhism
   conscious dying 144

C
Cabal, the
   another name for the Repterran Controllers 89
Cannon, Dolores 12
Carlsberg, Kim, The Art of Close Encounters 70
Carman, Elizabeth and Neil
   Cosmic Cradle - Souls Waiting in the Wings for Birth, ISBN 1-887472-71-1. 48
Carroll, Judy v, vii
   as star seed 186
   Human by Day, Zeta by Night- A Dramatic Account of Greys Incarnating as Humans ix
   The Zeta Message- Connecting All Beings in Oneness ix, 29, 108
Carter, Rev. Michael J. S. i
   Alien Scriptures- Extraterrestrials in the Holy Bible 181
Ceto, Sanni
   see Sanni Ceto 64
Charles, R.H.
   The Book of Enoch, tinyurl.com/pe9rk9 52
Churchill, Winston 179
Clinton, William 90
Coleman, Graham (Editor) et al
   Tibetan Book of Living and Dying, http //amzn.to/2iRsy0t 142

NAMES AND WORKS

conscious dying 144
Constantine
    Controller Emperor who removed reincarnation from the Bible. 58
Controllers 130
Cook, Patrick 181
Copernicus
    suppressed by Controllers 59
Corso, Col. Philip 160
Crissey, Brian L., Ph.D. ix
    NotMadeByHands.com
        bit.ly/2Ajzqb8 223
Crivelli, Carlo
    Annunciation with Saint Emidius 183
Crop Circle Connector
    bit.ly/2j4FsYG 223

D

Daly, Timothy J. 181
Darwin, Charles 136
Davis, Dr. Bob 182
Duran, Krsanna
    Web of Life and Cosmos- Human and Bigfoot Star Ancestors v

E

Eastman, Mark 181
Egarina- now Valerie Barrow, wife of Alchquarina, Lyran Commander-in-Chief of the Rexegena 45
Egyptian, derived from Gypsy 55
Einstein, Albert 136
El Or Kah 144
    reincarnated as Judy Carroll 61

Emoto, Masaru
    The Hidden Messages in Water
        bit.ly/1cwk43y 199

G

Gaddis-Cowles, Vincent & Margaret
    The Strange World of Animals and Pets, 1970. 70
Gantei
    A mysterious image of a white horse led this priest named to climb Mt. Kurama in 770 AD, which led to discovery of Reiki. 56
Ghirlandaio, Domenico
    Madonna with Saint Giovannino 183
Goswami, Amit, Ph.D,
    Physics of the Soul (2001) 223
    The Self-Aware Universe- How Consciousness Creates the Material World (1993) 223
Greer, Steven, Ph.D.
    "Unacknowledged
        An Expose of the World's Greatest Secret" amzn.to/2CvtfpP 198

H

Hambleton, Steve, Ph.D. 154
Hansen, Suzy
    The Dual Soul Connection
        The Alien Agenda for Human Advancement 202
Hellyer, Hon. Paul 75, 89
Hernandez, Reinerio 182
Hoffman, Dale Allen
    bit.ly/2j0pGen 126

I
Isis
 Set's twin sister and wife of Osiris 55

J
Jesus
 Guardian who attempted to reverse Reptilian control. 56
 suppressed by Controllers 59
John the Divine 55
Joseph of Arimathea 55
Juju
 see Kuita,Kujuolu
Jung, Carl
 sensing, feeling, thinking and intuiting 153
Justinian
 Controller Emperor who removed reincarnation from the Bible. 58

K
Kaku, Michio, Ph.D.
 Hyperspace
  A Scientific Odyssey through Parallel Universes, Time Warps, and the Tenth Dimension. (1994) 224
Kaye, Helene 25, 108, 113, 142
 aura cleanse on YouTube channel zetaguardian1 142
 daughter Kira 25
 Ka Li Yah 12
Kelle, Dr. Raymond Andrew, Ph.D.
 Venus Rising- A Concise History of the Second Planet v

King, Martin Luther, Jr.
 suppressed by Controllers 59
Klimo, Dr. Jon 182
Krishna
 Guardian who attempted to reverse Reptilian control. 56
Kuita,Kujuolui 208

L
Lipton, Dr. Bruce
 Spontaneous Evolution, with Steve Bhaerman 34

M
Mackenzie,Vickie
 A Cave in the Snow, 97
Mao-son, from Venus 56
Maris 33, 112, 212
Mary
 Jacobe 55
 Magdalene 55
 the Mother of Jesus 56
McClendon, Sarah 90
Mendonça, Miguel
 Being With the Beings 208
 Meeting the Hybrids 208
Missler, Chuck 181
Mitchell, Astronaut Edgar 182
Moon, Peter
 Transylvanian Moonrise 55
Moses 79

N
New Humans 49
Nina 114
NotMadeByHands.com
 bit.ly/2Ajzqb8 223

O
O'Grady, Aingeal Rose
 The Nature of Reality and others amzn.to/2jaDYMm 199

Ocean, Joan
    Dolphins Into the Future 191
Oris 30, 65, 212
    a tall Grey Elder Teacher 25
    Aura-cleansing technique, bit.ly/2iBmvK4 142

P

Palmo, Tenzin 97
Parrott, Brett
    bit.ly/2j4FsYG 223
Pearl, Dr. Eric 35

R

Radin, Dean, Ph.D,
    The Conscious Universe (1997) 224
Rockefeller, Jean i
Rodwell, Mary i
Rosin, Dr. Carol 90
Rosin, Hannah
    The End of Men vi
Rumi 130
Russell, Peter
    From Science to God. (2002) 224

S

Saint James, the brother of Jesus 55
Sanni Ceto 67, 68, 109
    Stranded on Earth 10, 65
    Zeti Child 10, 65
Schild, Dr. Rudy 182
Schwartz, Gary, Ph.D., and Linda Russek, Ph.D,
    The Living Energy Universe. (1992) Schwartz and colleagues (2001) show evidence of life beyond human existence in"After Life Experiments." 224
Schwartz, Gary, Ph.D., et al
    After Life Experiments. (2001) 224
Serapis 212
Sitchin, Zecharia v
    bit.ly/1PAOUOB 47
    The 12th Planet and other books in the Earth Chronicles series. 47
Smith, Jacquelin
    Animal Communication- Our Sacred Connection, and Star Origins and Wisdom of Animals- Talks With Animal Souls. 11
    reincarnation of Zan Tu Kai 223
Spiers, Aunty Beve 114
Sprinkle, Dr. R. Leo vii, 35
    "ET Experiencers- From Planetary Persons to Cosmic Citizens?" 34, 223
Stiffelman, Susan
    Parenting Without Power Struggles 186
Stoner, Chris 91
SunBow, as told by Sasquatch Elder Kamooh
    The Sasquatch Message to Humanity Book 1 (2016) amzn.to/2EPcXWY. Book 2 (2017) amzn.to/2lPiRyQ. 225
Suzy Hansen
    The Dual Soul Connection- The Alien Agenda for Human Advancement 224

Swanson, Claude, Ph.D.
    Life Force, the Scientific Basis
        Vol. 2 of the Synchronized
            Universe (2011) 224
    The Synchronized Universe
        New Science of the Para-
        normal. (2008) 224
Sydney 113

T
Tart, Charles T., Ph.D. 224

U
Usui Mikao, founder of Reiki 57

V
Virgil
    Aeneid amzn.to/2jmz2kY 198
von Braun, Werner 90

W
Wilde, Sherry
    The Forgotten Promise 225
Wolf, Dr. Michael 75, 89, 91
Wolf, Fred Alan
    The Body Quantum 71

Y
Yeshua, see Jesus 126

Z
Zan Tu Kai 61, 144

Index

Numerics
11 Spiritual Laws 171
11 Universal Laws 171
25,625-year cycle 35
5,125- year cycle 35
9/11 144

A
abductions 93
  by Controllers 73, 87
Abel
  symbolizes the higher path of unconditional love 50, 129
Aborigines ii, 46, 107, 111, 114, 120, 131
  as Wisdom Keepers and Caretakers of the Land 131
  creation myth 114
  death process 141
  dreaming misinterpreted 131
  Dreamtime 48
  taboo saying the name of the deceased 142
  Wise Woman 114
acceptance 85, 147
ACERN i
Adam
  as a mutation taken from "Eve's rib" 123
  being created out of "clay" 123
Adam and Eve 49
  parable 123
  symbolic of opposite polarities 124
  true story of 46
addictions 158
additives 186
advice
  Always feel a message in your heart and discard any information that feels unloving, hurtful or fearful. 81
  Be at peace in your heart so pure love can flow forth. 163
  Choose to act for the good of all. 194
  Direct your energy into positive and loving thought patterns. 194
  It is only through love that the long and sometimes difficult journey that all of us are undertaking can be satisfactorily completed. 180
  Learn to focus more upon the similarities rather than upon the differences between yourselves and others. 163
  Learn to think for yourself and to discriminate between pure and impure emotion. 178
  Love is your power, your strength and your guidance. 164
  Simplify your life so you are not overwhelmed by the complexity and confusion that takes up so much valuable time and energy, stopping you from putting time aside for spiritual nourishment. 193
  Stop judging yourself. 164
  Tap into Source, and your vision will be clear, your heart will be filled with love, and your mind will be free to create endless possibilities. 165
  The power of love conquers everything in its path. 166
  To evolve to higher levels empower yourself 158
  We attract to us what we need to learn or heal within self, on multidimensional levels. 164

You have the power to walk straight through fear unscathed. 165
afterlife 148
  manifesting one's expectations 140
Aikido 21
Akashic Records 57
  affected by Controllers 76
Alcheringa 44, 63
Alchquarina 111
aliens 163
androids 63
Angel of Light 81
angels 29, 110, 143
  image of 82
anger 164, 178, 179, 187
  and empowerment 177
  exploitation of 177
  how to handle 177
  normal 177
animal kingdom 155
  and emotions 179
  herd instinct 161
animal mutilations
  and Controllers 87
Ant People 209
  positive, mainly based on the moon. 210
Antares star system 64
Anunnakis 52, 74
  not all Anunnaki people on Earth are Controllers, not all are negative 79
appearances
  used to deceive 82
Aramaic 126
  Prayer of Yeshua 126
Archangel Michael 81
Arcturians 62
  blue-grey 61
  crop circles 199
  egalitarian, gender neutral 53
  life span 61
Arcturus 61
artificial insemination. 50
Aryan Ideal 76
Ascended Masters 81, 88
  and Controllers 81
ascension 96, 147
  assisted ascension program 185
  distorted beliefs about 140
  fake news about 99
  key to 149
  responsibility for our own 147
assistant creators 17, 44, 57, 123, 124
astral plane 22, 53, 58, 70, 140, 156
  after death 96
  and the Repterran Controllers 57
  as a barrier or mirror 37
  as a common meeting ground for helpers to connect with humans who has chosen to evolve 38
  attack by lower astral entities 23
  chaotic from the fear and superstition created in human minds by the Repterran Controllers 57
  chaotic, with many conflicting thought forms 22
  clearing of fear 37
  dreams, illusion and emotion 24
  illusion after death 140
  illusion screens from belief systems 96
  illusion traps 59
  interference mistaken for ET contact 23
  key to freeing yourself from its illusions 24
  known as the Plane of Dreams, Emotion and Illusion 57
  safe experimentation 156
  trickery 156
astral travel
  during the sleep state 152
Atlantis 52, 53, 107
  demise 53, 132
  energy transmutation 128
  Repterran use of crystals to reprogram people 132
  rise as spiritual reincarnation 100
  unspiritual 57
atomic bomb 98
aura 110

Index

cleansing technique to use in conjunction with Reiki 142
aura cleanse
  see YouTube zetaguardian1 channel 142
Australia 107, 113, 131
  oldest country on Earth 46
  part of Lemuria 107
  site of Rexegena survivors 45
Australian Bush Flower therapy 115
Ayers Rock 47, 108

B

bacterial meningitis 69
balance 157
Ballarat 121
baptism
  represents cleansing by water 133
Bearded Dragon lizard 116
belief system 85
Betelgeuse star system 61
between-life choices and decisions 172
Bible 79
  "Judge not or you yourselves will be judged." 147
  "sins of the father will be visited upon the son" reference to karma and reincarnation 146
  "The Kingdom of Heaven is within." 34
  "reap as we have sown," reference to karma, or balance 146
  Anti-Christ as Being of Light 84
  deliverance is by grace only, Ephesians 2 147
  Eve taken from Adam's rib 123
  Fall of Man 49, 129
  fallen "Watchers" took the "daughters of men" as wives. 50
  fallen angels 69
  Genesis 72, 123
  go forth and multiply 124
  Great Flood and demise of Atlantis 53
  Humans created in the image of God 52
  Jesus tempted by Satan 23
  John 14.2 18
  love your neighbor 147
  Lucifer as the Repterrans 50
  many mansions as dimensional aspects of the multiverse 137
  Noah and his family 130
  practice good works, James 2 147
  temptation of Eve in the Garden of Eden 49
  Ten Commandments from Repterrans 79
  Tower of Babel example of divide and conquer 79
  written by patriarchal Repterrans 133
bigotry 161
Birthing Stone 120, 121
Blond ETs 209
Blonds 55, 88
  deceptive practices 82
Bohemia 45
brain parasite 67, 130
  effect on mammalian humans 68
bridge
  between science and religion 136, 137
Buddha
  Consciousness 129, 147
Buddhism 96
  closest to higher off-planet spiritual systems 54
bullying 161

C

Cabal
  see Repterrans and Controllers 89
Cabal, see Repterrans and Controllers 89
Cain
  killing Abel, which is an allegory representing the lower path

that was chosen by the New
Humans 50, 129
symbolic of the lower path of fear
(jealousy) 50
calligraphy 158
Canada 74
cancer cure
Rife vi
Carroll, Judy
as assistant creator 17
memory of soul retrieval as a
Grey 144
cave paintings 54
chakras 35, 114, 132
all seven must be operating in
balance for good health. 179
sensitive to music 155
third-eye 50
channeling 81, 88
authentic is always uplifting and
empowering 82
red flags 84
chariots of fire 51
Charismatic Christianity 80, 84
and criminal activities 84
people trapped by 83
Charismatic Christians
use of chi energy 21
Children
Crystal vii
Indigo vii
Rainbow vii
Star vii
choice 194
free-will 193
Christ 148
Consciousness 129, 147
life as an allegory 148
clairvoyance 50
clones 209
cloning 205
co-creators 17
collective consciousness 195
commandments 151
common ancestor
plants and animals 71
communism 90
compassion 147, 155, 192

concentration
Earth children losing ability 157
confidence games 85
confusion 163
connection 112, 164
conscious awareness 99
conscious dying 96, 140, 144
consciousness 137
and spiritual mystery 137
collective 191
expanded 137
consciousness-enhanced
technology 210
contactees
and poltergeists 23
control 18, 148
Controllers 26, 53, 55, 59, 73, 79, 93,
124, 133, 139, 143, 147, 149, 150,
185, 188, 209, 211
abductions 73, 87
agenda 140
aim to block human evolution 74
amnesia nets 140
and fear 99
and fundamentalist religions 73
and New Age beliefs 80
and psychics 84
and royal bloodlines 79
crop circles 198
disguised as business, military,
government and church
leaders, law makers, etc. 86
distorting truth to remain in
control 73
exploit fear and anger 178
feed off human fear and
emotion 82
focused on perfection of the
physical Earth-human
form 76
infiltration of society 76
influence can be overcome by
evolving to higher levels of
mind/spirit 87
influence on astral plane 76
influence over many Outer Earth
authorities and
organizations. 86

235

Index

interference 73, 154
motivation 91
refuse to relinquish their ownership and complete control of Earth and humanity. 53
removal of reincarnation from Church doctrine 141
Roman Emperors 58
shape-shifters 82
some genuinely care for the welfare of humanity and Earth 79
spread the false belief of a judgmental God figure seated on a heavenly throne, condemning sinners to an eternity of hellfire and damnation 139
suppressing spiritual growth 75
trapped in Earth-human form 148
using humans 73
cosmic citizenship 207
earning our right to 98
Cosmic Law 44
cosmic racism 197
Crabwood formation 198
creationism 136
and the supernatural 136
Creator Beings 16, 98, 109, 110, 111, 114, 115
driven from Earth by the Repterrans 112
Creator Beings, see also Elohim 195
cremation
and dogma 139
Cro-Magnons 209
crop circles 30
messages 198
Milk Hill 29, 33
multidimensional 33
subtle energy encapsulated within 124
Trojan horse 198
Crystal Children 185
crystals

absorb and magnify positive or negative energy 132
energy dangers 131
for healing 132
have consciousness 132
cult busting 84
cynicism 187
Czechoslovakia 45

D

dark 150
Dark Ages 58
death 139
and anger, fear, pain and frustration 143
letting go of the deceased so their spirit is not trapped 141
nobody is ever left to die alone and unaided 144
deceit 83, 192
December 21, 2012 81, 99
deception 82, 84
DeGelder, Aert ii
déjà vu 117
demons 74, 85, 181, 182
De-raks 67
aggressive 67
as shape shifters controlling Earth authorities 67
description 67
derision 187
devolution 74, 124, 155, 157
back from human to animal. 130
dinosaurs 73
discernment 84, 212
disclosure 90
disempowerment 125
dishonesty 192
disinformation i, 22, 73, 83, 93, 97, 99, 139, 143, 198
campaign, purpose of 74
example of Greys "stealing people's souls." 143
distractions 157
and operating ET ships 157
distrust 145
DNA 109, 113, 185
"junk" 35

enhancement 49
dogma 151
 blocking spiritual progress 139
Dorado Sector 66
Dr. Edgar Mitchell Foundation for Research into Extraterrestrial Encounters (FREE) 210
Draconians 44, 45, 66, 189
 and Repterrans 63
 as fallen angels 69
 as fallen race 68
 as source of legends of Satan 68
 brain parasite 68
 creators of the Annunaki 47
 once quite highly evolved 69
 rocket science and space travel 51
 seen as gods by New Humans 51
 trapped on Earth by ego, fear and need to control 68
drama
 addiction to 82
Dreamtime
 firmly based on the story of Ancestor/Creator Beings who came to Earth from off-planet 48
drug culture 81
drugs
 psychotropic 83
dual soul
 confirm each other's memories 203
duality 124

E

E. coli
 age 68
 asexual binary fission 68
 DNA 69
Earth
 heavy atmosphere compared to Meropes in the Pleiades 45
 hijacking of 45, 189
 known widely as a war planet 125
 Level One planet 63
 magnetic field reversal 36
 moving towards level two 64
 multidimensional 86
 number-one "school planet" in the universe 125
 one of the lowest-frequency planets 68
 spiritually quarantined 125
 very much a "school planet" 205
Earth civilizations
 Atlanteans vi
 Controllers vi
 Egyptians vi
 Lemuria vi
Earth humans 53
 carrying a higher percentage of Repterran, Anunnaki and Draconian bloodlines 80
 self-centered 191
Earth-human parent
 brought on board ship to bond with the hybrid child 205
Earth-plane humanity
 out of sync with the rest of the universe 68
egotism 23, 24, 37, 59, 84, 129, 148, 151, 153, 161, 163, 164, 193, 194
 attracts thought forms 24
 none beyond astral plane 24
Egypt 139
 and Kariong 121
 belief systems Controller-based 58
 Guardians as priests and priestesses 54
 mummification 139
 mummification trapping souls 58
 Mystery Schools 58
 story of Osiris, Isis and Set 55
Egyptian mystery schools. 79
El Or Kah ix
electromagnetic field
 prevents dematerialization 91
Elites 73
Elohim and the Controllers 73
emotions
 in Greys 176
empathy 146, 155
enemies 85
energy 136
 free vi

INDEX

vs. matter 17
Enoch, Book of, see Charles,
    R.H. 88
envy 161
Essassanis
  hybrids 205
  operate on a physical level 202
ET
  ancestors i
  contact removes certain negative
    blockages from the aura 23
  craft ii
  food requirements 202
  genetics 188
  in art ii
  intervention, prerequisites
    for 132
  technology 75
  visitation increasing 92
ET civilizations ix
  Arcturians vi, 16
  Draconians vi, 16
  Grey Guardian teachers 16
  Martians vi
  Pleiadians vi, 16
  Sirians 16
  Venusians
    Jesus vi
    Tesla vi
  Zeta Reticulans (Zetas) 16
ET contact 109
  accelerating spiritual growth 182
etheric energy
  chi/ki/qi 21
  for self-defense 21
ETs
  examples of peace, tolerance,
    respect for the environment
    and unconditional love 189
  incarnating as humans 189
  increasing rate of incarnation as
    humans 189
  main purpose for coming is to
    help you to open conscious
    awareness to higher
    realities 99
  mission 189
  not to be worshiped 189

Eucalyptus 115
evil 161
evolution 71, 74, 88, 95, 98, 99, 136,
    150, 152, 188, 195
  from animal human to cosmic
    human by choosing love over
    fear, thus ending war 98
  reasoning power 49
  spiritual 96, 136
evolutionary healing 36
exoskeleton 197
expanded conscious awareness 202
exploitation 148
extinction 67

F
failure 193
fake news
  "Pleiadian" anti-Grey
    propaganda 83
  accusations against Greys 143
  as we evolve spiritually our
    lower chakras gradually
    close down 179
  ascension to happen suddenly 99
  autism in children blamed on
    negative interference by the
    Greys and their hybrid
    program 186
  denying the existence of any off-
    planet human 73
  earliest instance 45
  Earth humans alone were created
    to look like God 74
  ETs are demons 181
  Greys are evil 83
  labeling all ET contact as
    demonic 73
  Roswell crashes were
    accidents 172
  speaking in tongues 80
  stories of negative gray and
    reptilian aliens who are here
    to abduct people or take over
    Earth 93
  you are a poor, helpless sinful
    little being 151
fall of man 49, 130

family
   off-planet and Earth-plane 211
fear ii, 17, 22, 23, 24, 27, 37, 58, 59, 65, 73, 82, 86, 93, 97, 129, 140, 145, 149, 160, 163, 164, 178, 181, 182, 186, 189, 197, 198
   and antidote to counteract it 165
   -based belief systems 99
   blockages 96
   feeds off fear 165
   from the past and present 162
   how to handle 178
   keeps us safe 178
   main issue on Earth 207
   necessary as protection 178
   normal 177
   of birthing our consciousness into the new age 162
   of Grey's appearance 161
   taking on others' fear as your own 165
Fifth World 163
   consciousness 193
first humans on Earth 46
forbidden fruit 128
France
   Camargue region 55
   Les Saintes-Maries-de-la-Mer 56
FREE i, 182
   and hypnosis 182
   see Dr. Edgar Mitchell Foundation for Research into Extraterrestrial Encounter 210
free will 49, 50, 93, 95, 97, 98, 128, 130, 132, 193, 194, 205
   as protector against ET intervention 132
   dangerous without self-discipline 129
   not as much as Earthlings like to think 25
   vs.God's will. 194
free-will choice between love and fear 88
Fundamentalism 83, 124
future
   made up of possibilities, never certainties 82

G

Galileo
   suppressed by Controllers 59
Gandhi, Mahatma
   suppressed by Controllers 59
Gantei 56
Garden of Eden v, 46, 123, 124, 130, 133, 156
   now Australia 131
   serpent 128
genetic engineering 49
genetically upgraded children 185
Gobi Desert 54
God 16, 129, 137, 146, 161, 172
   created everything 149
   different paths to 147
   eternal 146
   gender vi
   nature of vi
   not a person 79
   of the Old Testament
      prone to human weaknesses of jealousy, anger and the need for praise and flattery 79
      was the Repterrans 79
   Source Energy of the universe 149
   Spark of, within 151
   will 194
Gods of Olympus
   Repterrans 53
Grandmother Tree 116
gratitude 112, 193
Greater Reality 141
   not physical, but a holographic energy whose projection forms what we mistakenly perceive as "reality" on Earth. 141
Greece 203
greed 129, 161, 163
Greer, Steven 210
Grey
   tall Greys are of a higher, less physically oriented frequency 202
Grey Elders 210
Grey eyes

mirrors to the Earth-human
    subconsciousness 162
Greys i, 62, 64, 144, 172, 176
  and emotions 176
  as "triers" and "testers" of
      souls 37, 95
  as insectoid humans 176
  as master gem cutters and
      polishers of the universe 189
  cannot chew food 203
  celebrate with dance, song and
      other festivities. 176
  Councils of Elders 194
  emissaries 61
  experience hope, joy and
      sadness. 180
  explorers 61
  eyes reflect or mirror buried
      fears 37
  gender differences 61
  have a very well developed sense
      of humor, 176
  have great multicultural "parties"
      up on the ship, as a way of
      coping. 176
  laugh and cry 180
  major function, assisting in
      emergencies like 9/11 144
  major function, heavy soul
      retrieval 143
  marriage ceremony 204
  some are related to dolphins 65
  speaking out as Earth ETs 113
  telepathic 61
  totally egalitarian 203
  tough love from 158
  understand emotion at a
      different energy frequency
      that is not readily perceptible
      by Earth-humans. 180
  warnings about 83
  waste matter is cleared through
      the skin or breath 203
Greys' appointed task
  to assist you past the limitations
      imposed upon your minds
      by fear 162
Group Consciousness 194

Group Soul Consciousness 191
Guardians 58, 75, 86, 92, 93, 95, 97, 98,
    123, 130, 144
guides
  do not label themselves 24
guilt 140, 193
gurus 163
Gypsies 55
  carrying old teachings 55
Gypsy Guardians 56

H

halo 110
harmony 157, 166, 208
hatred 161, 163
Hawkins, Dr. David R., M.D.,
    Ph.D. vi
Haystack Air Force Laboratory 91
hierarchies 29
hieroglyphs 46, 54
Hinduism
  and Buddhism 54
History
  of ET Contact on Earth 103
history ix
hive mentality 191, 195
holding onto the person's ashes
  effect of 141
holographic projection of our
    perceived reality 141
hominid body shape 52
Homo sapiens sapiens 51
  descended from Pleiadian/
      Lyran/human offspring
    Human origins 46
  Pleiadian DNA 47
human beings
  all experience emotion 176
human kingdom 155
Human Ladder 22, 29, 30, 33, 35, 69,
    74, 95, 96, 99, 113, 124, 125,
    129, 155, 185, 189, 191, 195
  10 levels 25, 159
  and balanced emotions 178
  from Earth human to cosmic
      human. 29
  from human to angel 29
  Level Four 28

first step into the Guardian realm 28
  Level One 34
  Level Three 27, 28
  Level Two 26
  not a physical hierarchy 25
  one-directional movement in 26
  reflects ascending frequencies, nothing to do with "good" or "bad." 25
human species
  non-primate types 74
humanity's call for help 98
Humans
  1,000 other Level-one human species 25
  all have off-planet heritage 79
  and God Consciousness 70
  as caretakers 179
  as reasoners 70
  developed from many different animal species 70
  greatest challenge is the expansion of conscious awareness 99
  losing the sense of wonder and imagination 187
  more and more out of balance and harmony 187
hybrids
  "mixture" creatures 46
    captured for reproduction 46
  energy frequency 46
  present-day ET hybrid program 46
  primate/reptoid/Pleiadian offspring 46
  unnatural 53
hypnosis 182

I

Illuminati 73
illusion nets 140
illusion of separation 146
imagination
  dismissed 188
immortality 128
implants 209
incarnation
  of ETs as humans 93
  purpose of 26
India
  Guardian base 58
Indigo Children 185
individuality 191
initiation 23
Inner Earth 53, 70
  Controller bases 76
  Zeta/Pleiadian base 54
insectoid humans
  cannot talk 65
  use telepathy 65
insectoid people
  have lightweight bodies designed for intergalactic space travel so the G forces won't harm them 65
  not as large in stature as most humanoids and reptoids 65
Insectoids 210
insecurities 162
instinct 49, 128
intelligences i
Intelligent Design 137, 223
interference 207
intervention 44
Isis 55
Italy
  Torino crop circle 198

J

jealousy 129, 161, 163
Jehovah vi
Jesus ii, vi, 34, 83, 126, 137, 140
  apostles 80
  Controller corruption of teachings 58
  death and resurrection symbolize the evolution of the spirit 140
  tempted by Satan 23
joyfulness 187
judgment 139, 140, 147, 163, 164
judgmentalism 161
Judo 21
JuJu
  see Kuita, Jujuolui

INDEX

K

Kamooh, Sasquatch Elder 18, 40, 41, 42, 51, 58, 59, 225
Karate 21
Kariong 114
  glyphs 114
karma 132, 140
  wheel of 96
Kata Tjuta 108, 110
Kaye, Helene
  as reincarnated artist 114
Kebbans 65
Key Creator- God 17
Krishna
  Consciousness 129, 147
Kuita, Jujuolui 208
Kung Fu 21

L

labels 24, 151
lack 193
Latin 198
laws of physics
  do not always apply elsewhere 153
Lemuria 107, 113
Level-two reality
  Everyone's needs are met without poverty, hunger or misery 27
  longer life spans 27
  more peaceful 27
life spans
  being cut short because of an incompatibility with the changing planetary energies 162
  longer in adepts 27
Light 18
like attracts like 22
Lizzies 208
Lord's Prayer 126
love 17, 112, 147, 149, 164, 177, 192, 208
  is a much higher frequency than fear, and is unlimited. 165
  is your power 151
  unconditional 155

Lucifer 69
  as Light Bringer 50
Lucy 44
  original Eve? 47
lust 129
Lyrans 110
  record keepers 62
  reproduction of 62
  seed colonies 62
  upgrading the slave species to a higher vibrational frequency 48
  vegetarians 62

M

magic 51
Mahabharata 79
Mantis 65, 210, 211
  as Elders to the Greys 212
  beings seven or eight feet tall. 65
  people 65
mantra 165
many mansions 137
Mars 53
  Mt. Olympus, home of the Gods of Olympus (Repterrans) who escaped Atlantis 53
  pyramids and other structures correspond to similar structures on Earth 53
martial arts 21
  energy trick 21
masters 163
material possessions
  keeping souls Earth-bound 58, 139
materialism 73
May 25 celebration 56
Mayan Calendar 81
Mayans 110
media disinformation i
meditation 114
  healing for Roswell 174
memories
  erased 59
Men in Black 87, 211
Meropians 62, 63
  description 62

omnivores 63
meteor strike 67
meteorites 45
Middle East 54
military-industrial complex 90
Millennium Bug 81
mind control 76, 81, 185
mind-state at the time of death
   the controlling factor of what we experience between lives and in the next life 140
misinterpretations
   of wisdom teachings 99
MJ12 91
Moldavia 45
Moldavite
   emotional response to 45
moon
   as base for Ant People and Off-planet Greys 210
Mother Mary ii
Mother ships
   many different planetary beings may be on board at any given time 203
movies
   as Controller tools 154
   war and "negative-alien invasion" 154
Mt. Kurama 57
multidimensionality 38, 137, 146
musculature
   difficulty in expressing emotion as Earth humans do 180
music
   alcohol, drugs and sex 155
   alcohol-related lyrics 154
   heavy metal, Controllers' use of 155
   need to be very selective of the sort you expose yourself to 155
Mutitjulu Waterhole 111
Mystery Teachings
   off-planet origin 58

N
NASA 36
Native Americans ii
Natural and Cosmic Laws 99
Nazis 76, 90
   "bell" vehicle 210
Neanderthals 209
near-death experience (NDE) 148
need vs. want 172
neediness 163
negative 150
   beings 165
negativity 82, 193
   draws in more 193
New Age 81, 82, 88
new human species 124
New Humans 53, 128, 130, 132
   mating with other primates 130
   unusually long life-spans from off-planet DNA 131
new totalitarians 90
Nibiru 36
Noah's Ark
   symbolic of off-planet rescue mission 131
Noah's Flood
   as the demise of Atlantis 131
nonattachment
   vs.denial 178
Nordics 88
Northern Africa 54
NotMadeByHands.com 199
nuclear weapons 210

O
Occam's razor 184
off-planet war about 8,000 BCE 54
Olgas 108
Oneness 27, 28, 31, 38, 83, 97, 124, 136, 146, 149, 157, 161, 187, 189, 191, 195
   and off-planet lives 146
   lack of limits our spiritual maturity 193
   your only path back to Oneness is Love 141
oneness 129, 163, 176, 192, 208
Oneness/God 16
Operation Paperclip 90
orbs 28

INDEX

Original Sin 128
original sin 130, 133
   Eve misinterpreted 124
Orion Nebula 66
Orion Nebular Sector 66
Osiris
   being cut to pieces is symbolic of the Reticulan cultures destroyed by the Reptilians and Blonds 55
Out-of-Body Experiences (OOBE) 84, 148
   during sleep 27
overpopulation 205
Oxfordshire 120

P

pain
   releasing 164
panic 75, 173
parties 176
past lives 96, 107, 164, 188
   memories and the Fifth World 35
   memories of 148
   regression 96
paths "home"
   all are OK 151
patriarchy 58, 124
peace 27, 166
peer pressure 85
penis
   symbolic of Reticulan disempowerment and inability to reproduce 55
Pentecostals
   speaking in tongues 80
Photon Belt 36
physical form
   nothing but an illusion 161
physical plane
   lives like the characters in a stage production. 192
physical reality 202
Pictor Sector 66
pineal gland
   energy manipulation of 50
Pitjantjatjara 108, 109
Planet Earth

rescue mission, your part in the 207
Planet X 36
planetary shift 35
Play of Life 192
Pleiadian language 54
Pleiadians 45, 47, 55, 62, 81, 82, 109, 110
   egalitarian, gender neutral 53
   like cousins to Earth humans 62
   past lives as 83
   surrogate mothers 46
   upgrading the slave species to a higher vibrational frequency 48
   very closely related to Earth-humans 62
PLFs
   see Programmed Life Forms 210
polarities
   working together in Oneness 125
polarity 99
poltergeists
   and ET contactees 23
population control 205
positivity 193
   must be actively chosen 194
possessiveness 163
poverty 73
power 18
prayer vs. meditation 153
pre-birth agreements 37
pre-birth life plan 172
pre-flight preparation
   using Qi Gong 157
Prime Directive 199
Programmed Life Forms 210
Project Blue Book 92
psychic 50
psychics and channelers
   danger from ego or fear 24
Purpose of Earth -- Earth was created so that a wide range of beings with various frequencies could come together to meet and bless Earth. 18
Puy-yats 67

aggressive 67
description 67
pyramids 62

Q
quantum physics 137, 223

R
race, creed and culture
  as theatrical costume 193
racism 161
Rainbow Children 185
Rainbow Serpent 109
reality
  cosmic 162
  subjective rather than objective 202
reasoning 49, 98, 128, 130
re-birth choices and decisions 95
Red Center 108
Reformation
  positive intervention 59
Reiki 142, 209
  founding of 57
  master symbol 57
reincarnation 53, 58, 59, 68, 70, 95, 128, 140, 141, 146, 147, 160
  as changing soul containers 145
  intrinsic part of evolution 34
  non-acceptance as major block to further understanding and spiritual growth 160
  preparation for 192
  purpose 188
religion 24, 193
  exploited by Controllers 152
  initial purpose 151
  used by Controllers to promote their version of history 124
religions 82
  based on love 151
  distorted 73
  what is not OK 151
religious
  and spiritual beliefs i
  intolerance 161
  relics trapping souls of saints 141

Renaissance
  positive intervention 59
reprogramming 132
Repterrans 44, 45, 49, 66, 73, 74, 82, 124, 130, 133, 143, 150, 185, 186
  and crystal energy 132
  appearing as Greys to frighten people 86
  as a major part of the Atlantean culture, on which the Sumerian, Egyptian and other cultures were based 52
  as Annunaki 47
  as basis of Hitler's "Aryan Ideal." 52
  as owners of the planet and superior to the more modern Earth humans 70
  as the Gods of Olympus 52
  bestowing psychic and clairvoyant abilities on some Earth humans who are too spiritually immature to handle such abilities wisely 50
  blatant rebellion against a universal code of conduct 53
  caught up in their own power and ego 51
  caused themselves problems by hijacking Earth 52
  continuing to reincarnate on Earth in Earth-human form 53
  created by Draconians 63
  DNA interference by 129
  feel superior 79
  hijacked the Earth many millennia ago 68
  interfering to throw things out of balance 50
  like dictatorships 132
  millions of years older than Homo sapiens 51
  not all Repterrans on Earth are Controllers, not all are negative 79
  patriarchal 53
  takeover of Planet Earth 53

INDEX

threatened by the new and upgraded human species being developed. 49
use of divide-and-conquer techniques 51
weapons 53
worshiped by New Humans 51
Reptilians 63
  attack on the Zeta Reticulan culture 55
  higher cultures 63
  negative ones on Earth 208
Reptoids 45, 48, 209, 211
  genetics 188
  more evolved, here to help 74
respect 147, 163
responsibility 130
resurrection 140
Rexegina 44, 45, 54, 113, 117
  90 survivors of 45
  purpose of mission to Earth 45
Rife, Royal vi
Rigel Kantares star system 61
Rollright Stones 120
  energy essence from, see NotMadeByHands.com 120
Romanies 55
Roswell crash
  "accident" 172
  pre-planned at high level 173

S

sanctification
  requires many incarnations 147
Sanni Ceto 111, 209
Sasquatch 70
Sasquatch people
  created 50 million years before Homo sapiens. 44
Satan 68, 129, 163
  Jesus tempted by 23
Schumann Resonance
  changing 36
science 137, 153
screen images 197
scribbly gum tree 115
scribing 157
second genesis 73

secret societies 79
seekers 85
self-doubt 164
self-empowerment 164, 207
self-realization 98
self-righteousness 161
Set
  depicted as a crocodile 55
shamanism 83, 155
  "New Age" confined to the astral plane 155
  need to cleanse all levels 156
Shambhala
  base in Northern India 54
  destroyed by Controllers 54
shape shifters 82, 86
shelf beds
  mold themselves into the exact shape of your body 204
Sirians
  upgrading the slave species to a higher vibrational frequency 48
skepticism 187
Sky Fathers 48
slave races
  begun millions of years earlier by Draconians and their Repterran off-spring 47
slavery 209
slaves 86
  emotional and spiritual 93
Sonten 57
Sonten, shrine of 56
soul
  immortal 146
soul container ix
soul containers 17, 54, 63, 145
  use made of by Pleiadians and other humanoid ETs 63
soul retrieval 143
soul window 85
Source 85
Space Brothers 81, 88
  and Controllers 81
space travel 84
  possible interference with 63
speaking in tongues 80

246

spirit guides 188
spiritual
 quests 116
spiritual evolution
 blocked on astral plane 24
 limited by fear and ego 37
spiritual freedom
 key to humanity's 143
spiritual practice
 purpose of 152
spiritual progress
 what is blocking it 150
Spiritualism 81
spirituality 152
 the key to human evolution 152
Star Beings 48
Star Children
 accessing 20% of their potential consciousness 35
 and daydreams 187
 and imagination 186
 and medications 185
 easily manipulated by the Controllers 185
 exposure to negativity and violence from music and video games 186
 genetically upgraded 185
 Indigo/Crystal/Rainbow 185
 losing their natural sense of wonder and magic 187
 need for down time 186
 need to daydream 187
 prodigies 186
 protection via knowledge and understanding 186
 thrown out of harmony and balance by GMOs and additives 186
 very sensitive psychically 185
Star Nations 109, 172
star seeds 113, 186
Star Wars weaponry 93
stress 187
Summary 6
superconscious
 level of mind 96
supernatural
 nothing is 99
superstition 58, 73, 93, 99, 163
Sydney NSW, Australia 46

T

tall Greys 210
 in white robes 211
Tar-hig-gom
 description of 64
 life span 64
teacher
 mark of truly spiritual 163
teachers 54, 68, 147
technology
 advanced, used by Controllers 86
 back-engineering 210
 consciousness-enhanced 210
 handling wisely 157
 limitations 153
 prematurely endowed 157
telepathy 112, 192
Te-raks 67
 description 67
 subterranean 67
terraform 62
terrorists 90
Tesla, Nicola vi
The Forever- God 16
the only way "Home"
 to open your inner perception and higher awareness 99
The Planetary Alliance 64
three levels of meaning 123, 198
time 165
 not real 99
titles, genuine masters and angels do not claim 82, 84
Torres Strait Island 142
Tree of Knowledge
 reasoning power and free will 128
Tree of Life
 immortality 128
Trojan horse 198
trust 129
truth embargo i
Tutankhamun 121

INDEX

U
UFOlogists 153
UFOs
   understood in the cosmos 153
Uluru 47, 108, 110, 111, 112, 115
   taboo 108
unconditional love 129
Universal and Spiritual Laws of the Cosmos 97
Universal Law
   atomic weapons contrary to 98
   of Free Will 97
Universal Spirituality 170
universe
   created in God's image 137

V
Vatican 75
Vedic Teachings
   and Hinduism 54
veneration of God
   through laughter and joy 187
vibrational frequency 211
   of Earth is being raised and refined 162
victims of circumstance 172
virtual reality chair 204
vortex 120

W
Wandjinas 48, 63, 110
   Creator Beings 131
war 27
   driven by fear 161
Watchers 50
   description 52
water 18
   as a living, breathing being 18
Wiltshire County, England 29
Wisteria 116
World War II 89, 90, 92, 98
worry 193

X
xenophobia 182

Y
yin and yang 109, 110, 116

Z
Zen 157
zero-point magnetic field 36
Zeta Kebbans 210
Zeta Reticulans, see Zetas 28
Zeta Reticuli
   30 cultures 64
Zeta Soul Group 192
Zetas i, 62, 64, 107, 172
   and the Human Ladder 28
   angry at how Earth humans are treated 179
   angry at how the Earth is treated 179
   can no longer reproduce sexually 205
   egalitarian, gender neutral 53
   language 67
   live extremely simple and uncomplicated lives 157
   upgrading slave species 48

Wild Flower Press
Documenting The Unexpected
An Imprint Of
Granite Publishing
POB 1429
Columbus NC 28756

NotMadeByHands.com
Collecting, Decoding And Distributing
High-Life-Force
Cosmic Essences Since 2002

Granite-planet.net
The Hub Of The 5th World

Made in the USA
Middletown, DE
29 January 2018